RESIDENTIAL REMODELING & REPAIR

PROFESSIONAL REFERENCE

Paul Rosenberg

Created exclusively
for DEWALT by:

www.palpublications.com
1-800-246-2175

Titles Available From DEWALT

DEWALT Trade Reference Series

- Blueprint Reading Professional Reference
- Construction Professional Reference
- Construction Estimating Professional Reference
- Electrical Estimating Professional Reference
- Electrical Professional Reference
- HVAC/R Professional Reference – Master Edition
- Lighting & Maintenance Professional Reference
- Plumbing Professional Reference
- Residential Remodeling & Repair Professional Referen
- Spanish/English Construction Dictionary – Illustrated
- Wiring Diagrams Professional Reference

DEWALT Exam and Certification Series

- Construction Licensing Exam Guide
- Electrical Licensing Exam Guide
- HVAC Technician Certification Exam Guide
- Plumbing Licensing Exam Guide

For a complete list of The DEWALT Professional Trade Reference Series visit **www.palpublications.com**.

 Pal Publications, Inc., a licensee of DEWALT Industrial To
800 Heritage Drive, Suite 810, Pottstown, PA 19464, Tel.: 800-246-217

This Book Belongs To:

Name: ______________________________

Company: ______________________________

Title: ______________________________

Department: ______________________________

Company Address: ______________________________

____

____

Company Phone: ______________________________

Home Phone: ______________________________

Pal Publications, Inc.
800 Heritage Drive, Suite 810
Pottstown, PA 19464-3810

First edition published 2006

ISBN 0-9777183-1-X

10 09 08 07 06 5 4 3 2 1

Printed in the United States of America

A Note To Our Customers

We have manufactured this book to the highest quality standards possible. The cover is made of a flexible, durable and water-resistant material able to withstand the toughest on-the-job conditions. We also utilize the Otabind process which allows this book to lay flatter than traditional paperback books that tend to snap shut while in use.

Preface

There are no more common types of construction work than residential construction and residential remodeling. In no other area of construction are tradesmen more likely to crossover from one trade to another.

A special mix of reference information is required for residential and remodeling work including important information on everything from concrete and masonry to carpentry, walls and floors, roofing and siding, plumbing, HVAC, and electrical. I have also included a chapter on telephone installations, security (alarm) wiring, and cable TV wiring. Since these are now found in almost every home, I considered them necessary. I conclude with essential information on materials, tools, conversion factors, a glossary and a list of abbreviations.

Naturally, there may be some aspects of residential and remodeling work that I have not covered in sufficient depth for some readers. I will update this book on a continual basis and attempt to include material suggested by readers as well as keep pace with developments in the trades.

Best wishes,
Paul Rosenberg

CONTENTS

CHAPTER 1 – *Concrete and Masonry*. 1-1

CHAPTER 2 – *Carpentry* 2-1

CHAPTER 3 – *Walls and Floors* . . . 3-1

CHAPTER 7 – *HVAC* 7-1

CHAPTER 8 – *Electrical* 8-1

CHAPTER 10 – *Materials and Tools* . . 10-1

CHAPTER 11 – *Conversion Factors* . . 11-1

CHAPTER 12 – *Glossary* 12-1

CHAPTER 13 – *Abbreviations and Symbols* 13-1

CHAPTER 1
Concrete and Masonry

MATERIAL QUANTITIES PER CUBIC FOOT OF MORTAR				
	Quantities by Volume			
	ASTM Mortar Type and Proportions by Volume			
Material	M 1:¼:3	S 1:½:4½	N 1:1:6	O 1:2:9
Cement	0.333	0.222	0.167	0.111
Lime	0.083	0.111	0.167	0.222
Sand	1.000	1.000	1.000	1.000
	Quantities by Weight			
	ASTM Mortar Type and Proportions by Volume			
Material	M 1:¼:3	S 1:½:4½	N 1:1:6	O 1:2:9
Cement	31.33	20.89	15.67	10.44
Lime	3.33	4.44	6.67	8.89
Sand	80.00	80.00	80.00	80.00

STANDARD SIZES AND SHAPES OF CONCRETE BLOCKS

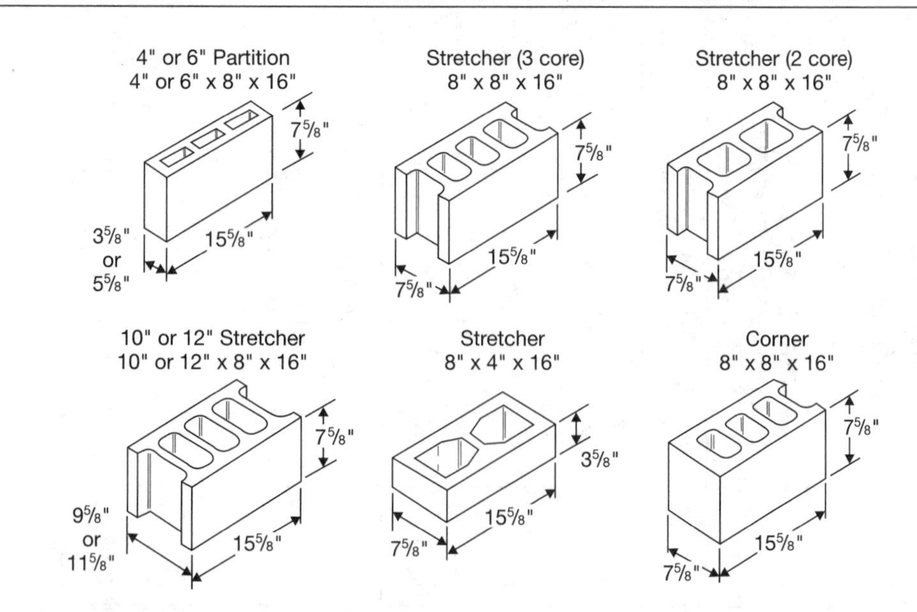

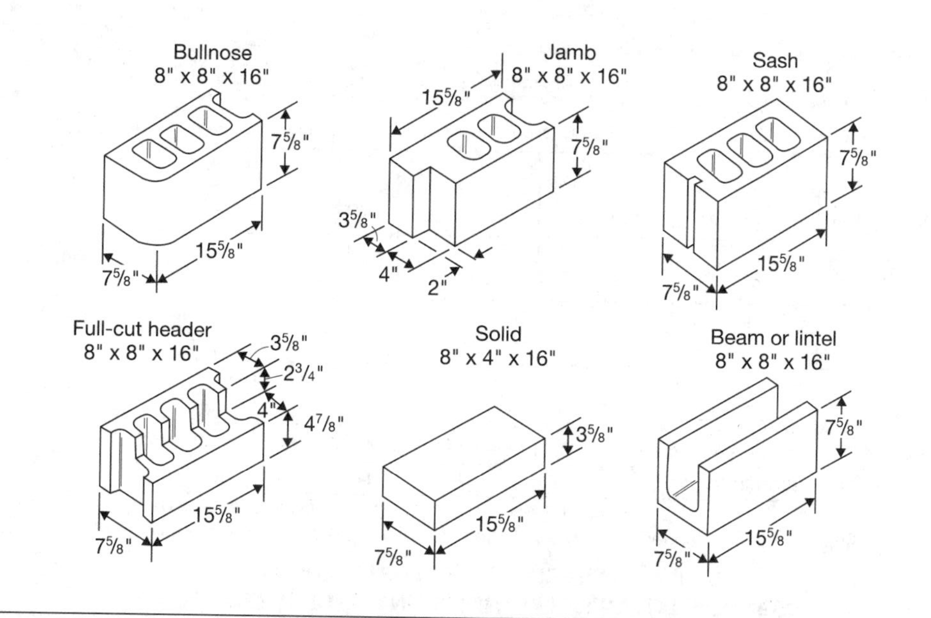

Bullnose
8" x 8" x 16"
Jamb
8" x 8" x 16"
Sash
8" x 8" x 16"
Full-cut header
8" x 8" x 16"
Solid
8" x 4" x 16"
Beam or lintel
8" x 8" x 16"

ALLOWABLE HEIGHT AND MINIMUM NOMINAL THICKNESS FOR CONCRETE MASONRY BEARING WALLS

For buildings up to three stories in height with walls of hollow or solid concrete masonry units. For buildings over three stories, see Building Code. (Thickness in inches)

Wall	Residential			Nonresidential		
	One-story	Two-story	Three-story	One-story	Two-story	Three-story
3rd story	—	—	8	—	—	12 8[d]
2nd story	—	8	8	—	12 8[d]	12
1st story	8 6[a]	8	8	12 8[d]	12	12 16[c]
Basement or foundation		12 or 8[b]		12 or 8[b]	12	12 or 16

[a]May be 6 inch for one-story single-family dwellings and one-story private garages when not more than 9 feet in height, with an allowance of 6 feet additional for any gable.

[b]When foundation wall does not extend more than 4 feet into the ground, the wall may be 8 inches thick. With special approval of the building official, this depth may be extended to 7 feet where soil conditions warrant such an extension. In no case shall the total height of 8 inch or 10 inch concrete masonry walls, including the foundation wall, exceed 35 feet.

[c]Must be at least 16 inches if the total height of the first, second, and third stories above the foundation wall, or from a girder or other intermediate supports, is more than 35 feet.

[d]Top story may be 8 inches when not more than 12 feet in height and roof beams are horizontal and total height of masonry wall is not more than 35 feet.

ALLOWABLE HEIGHT AND MINIMUM NOMINAL THICKNESS FOR CONCRETE MASONRY BEARING WALLS *(cont.)*

For buildings up to three stories in height with walls of hollow or solid concrete masonry units. For buildings over three stories, see Building Code. (Thickness in inches)

Wall	Cavity Residential			Cavity Nonresidential		
	One-story	Two-story	Three-story	One-story	Two-story	Three-story
3rd story	—	—	10	—	—	12 10[f]
2nd story	—	10	10	—	12 10[f]	12
1st story	10	10	12 or 10[e]	12 10[f]	12	12[g]
Basement or foundation	12, 10 or 8[h] (not cavity)			12 or 10 (not cavity)		

[e]In no case shall total height of a 10 inch cavity wall above the foundation wall exceed 25 feet.

[f]Top story may be 10 inches when not more than 12 feet in height and roof beams are horizontal and total weight of wall is not more than 35 feet.

[g]In no case shall total height of a cavity wall exceed 35 feet above the foundation wall, regardless of thickness.

[h]May be 8 inches for 1½-story single-family dwellings having a maximum height, including the gable, of not over 20 feet and having nominal 10 inch cavity walls. Such 8 inch foundation walls shall be corbelled to provide a bearing the full thickness of the wall above. Total projection not to exceed 2 inches with top course a full header course not higher than the bottom of the floor joists. Individual projections in corbelling shall not be more than one-third the height of the unit.

NUMBER OF CONCRETE BLOCKS BY LENGTH OF WALL

Length of Wall	No. of Units	Length of Wall	No. of Units	Length of Wall	No. of Units	Length of Wall	No. of Units	Length of Wall	No. of Units	Length of Wall	No. of Units
0' 8"	1/2	20' 8"	15 1/2	40' 8"	30 1/2	60' 8"	45 1/2	80' 8"	60 1/2	100' 8"	75 1/2
1' 4"	1	21' 4"	16	41' 4"	31	61' 4"	46	81' 4"	61	101' 4"	76
2' 0"	1 1/2	22' 0"	16 1/2	42' 0"	31 1/2	62' 0"	46 1/2	82' 0"	61 1/2	102' 0"	76 1/2
2' 8"	2	22' 8"	17	42' 8"	32	62' 8"	47	82' 8"	62	102' 8"	77
3' 4"	2 1/2	23' 4"	17 1/2	43' 4"	32 1/2	63' 4"	47 1/2	83' 4"	62 1/2	103' 4"	77 1/2
4' 0"	3	24' 0"	18	44' 0"	33	64' 0"	48	84' 0"	63	104' 0"	78
4' 8"	3 1/2	24' 8"	18 1/2	44' 8"	33 1/2	64' 8"	48 1/2	84' 8"	63 1/2	104' 8"	78 1/2
5' 4"	4	25' 4"	19	45' 4"	34	65' 4"	49	85' 4"	64	105' 4"	79
6' 0"	4 1/2	26' 0"	19 1/2	46' 0"	34 1/2	66' 0"	49 1/2	86' 0"	64 1/2	106' 0"	79 1/2
6' 8"	5	26' 8"	20	46' 8"	35	66' 8"	50	86' 8"	65	106' 8"	80
7' 4"	5 1/2	27' 4"	20 1/2	47' 4"	35 1/2	67' 4"	50 1/2	87' 4"	65 1/2	107' 4"	80 1/2
8' 0"	6	28' 0"	21	48' 0"	36	68' 0"	51	88' 0"	66	108' 0"	81
8' 8"	6 1/2	28' 8"	21 1/2	48' 8"	36 1/2	68' 8"	51 1/2	88' 8"	66 1/2	108' 8"	81 1/2
9' 4"	7	29' 4"	22	49' 4"	37	69' 4"	52	89' 4"	67	109' 4"	82
10' 0"	7 1/2	30' 0"	22 1/2	50' 0"	37 1/2	70' 0"	52 1/2	90' 0"	67 1/2	110' 0"	82 1/2

10' 8"	8	30' 8"	23	50' 8"	38	70' 8"	53	90' 8"	68	110' 8"	83
11' 4"	$8\frac{1}{2}$	31' 4"	$23\frac{1}{2}$	51' 4"	$38\frac{1}{2}$	71' 4"	$53\frac{1}{2}$	91' 4"	$68\frac{1}{2}$	111' 4"	$83\frac{1}{2}$
12' 0"	9	32' 0"	24	52' 0"	39	72' 0"	54	92' 0"	69	112' 0"	84
12' 8"	$9\frac{1}{2}$	32' 8"	$24\frac{1}{2}$	52' 8"	$39\frac{1}{2}$	72' 8"	$54\frac{1}{2}$	92' 8"	$69\frac{1}{2}$	112' 8"	$84\frac{1}{2}$
13' 4"	10	33' 4"	25	53' 4"	40	73' 4"	55	93' 4"	70	113' 4"	85
14' 0"	$10\frac{1}{2}$	34' 0"	$25\frac{1}{2}$	54' 0"	$40\frac{1}{2}$	74' 0"	$55\frac{1}{2}$	94' 0"	$70\frac{1}{2}$	114' 0"	$85\frac{1}{2}$
14' 8"	11	34' 8"	26	54' 8"	41	74' 8"	56	94' 8"	71	114' 8"	86
15' 4"	$11\frac{1}{2}$	35' 4"	$26\frac{1}{2}$	55' 4"	$41\frac{1}{2}$	75' 4"	$56\frac{1}{2}$	95' 4"	$71\frac{1}{2}$	115' 4"	$86\frac{1}{2}$
16' 0"	12	36' 0"	27	56' 0"	42	76' 0"	57	96' 0"	72	116' 0"	87
16' 8"	$12\frac{1}{2}$	36' 8"	$27\frac{1}{2}$	56' 8"	$42\frac{1}{2}$	76' 8"	$57\frac{1}{2}$	96' 8"	$72\frac{1}{2}$	116' 8"	$87\frac{1}{2}$
17' 4"	13	37' 4"	28	57' 4"	43	77' 4"	58	97' 4"	73	117' 4"	88
18' 0"	$13\frac{1}{2}$	38' 0"	$28\frac{1}{2}$	58' 0"	$43\frac{1}{2}$	78' 0"	$58\frac{1}{2}$	98' 0"	$73\frac{1}{2}$	118' 0"	$88\frac{1}{2}$
18' 8"	14	38' 8"	29	58' 8"	44	78' 8"	59	98' 8"	74	118' 8"	89
19' 4"	$14\frac{1}{2}$	39' 4"	$29\frac{1}{2}$	59' 4"	$44\frac{1}{2}$	79' 4"	$59\frac{1}{2}$	99' 4"	$74\frac{1}{2}$	119' 4"	$89\frac{1}{2}$
20' 0"	15	40' 0"	30	60' 0"	45	80' 0"	60	100' 0"	75	120' 0"	90

Note: Based on units $15\frac{5}{8}$ inches long and half units $7\frac{5}{8}$ inches long with $\frac{3}{8}$-inch thick head joints.

ANCHORING SILLS AND PLATES TO CONCRETE BLOCK WALLS

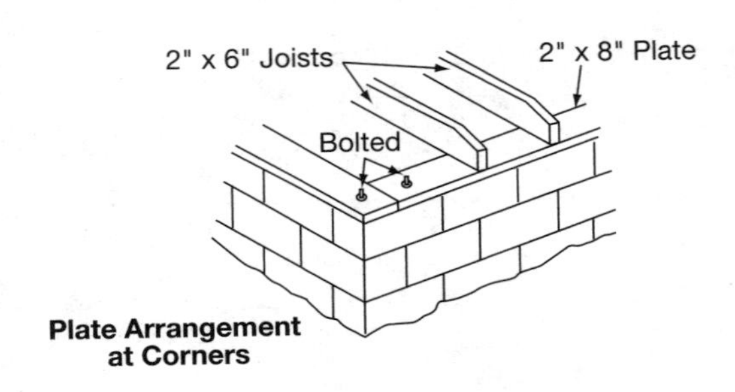

Plate Arrangement at Corners

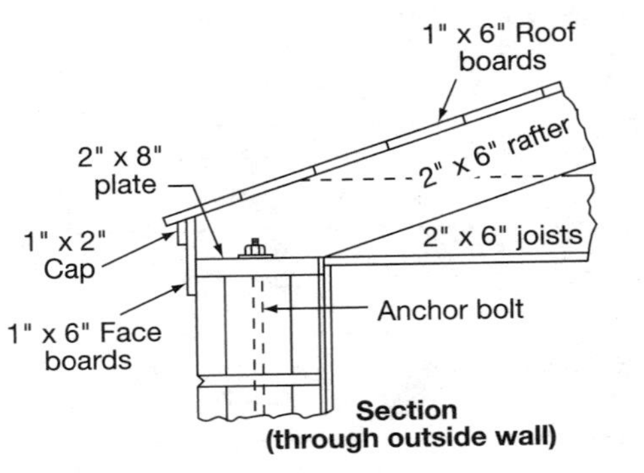

Section (through outside wall)

ANCHORING SILLS AND PLATES TO CONCRETE BLOCK WALLS *(cont.)*

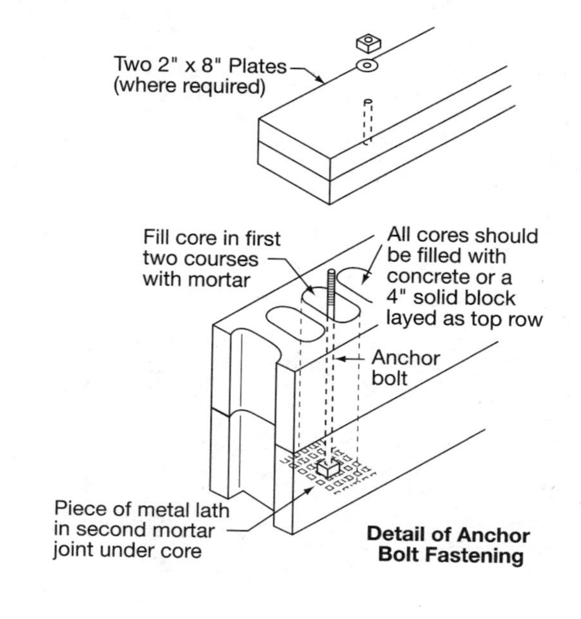

Detail of Anchor Bolt Fastening

INSTALLATION OF ELECTRICAL SWITCHES AND OUTLET BOXES IN CONCRETE BLOCK WALLS

INSTALLING VENTILATING AND HEATING DUCTS IN CONCRETE BLOCK WALLS

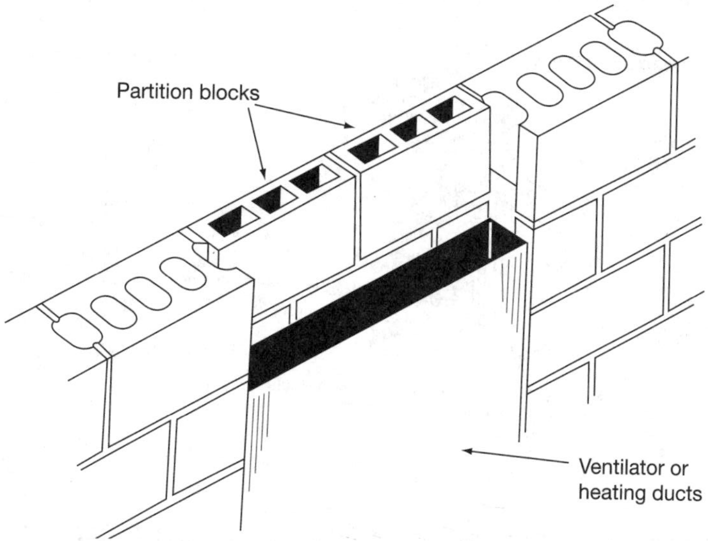

CROSS-SECTION END VIEW OF A SIMPLE BLOCK WALL

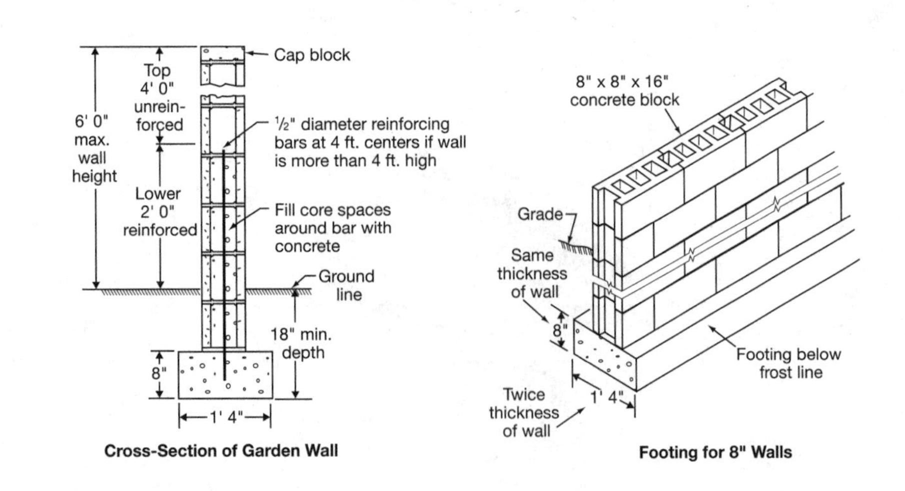

Cross-Section of Garden Wall

Footing for 8" Walls

Vertical reinforcement rods are placed in the hollow cores at various intervals.

HOW VARIOUS BRICKS ARE USED IN OVERLAP BONDING

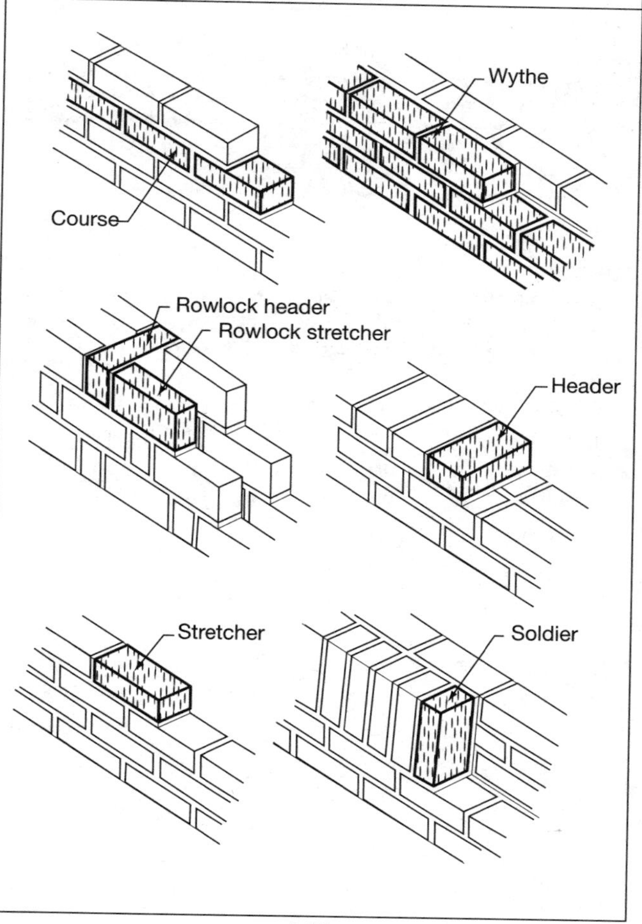

AMOUNT OF BRICK AND MORTAR NEEDED FOR VARIOUS WALL SIZES

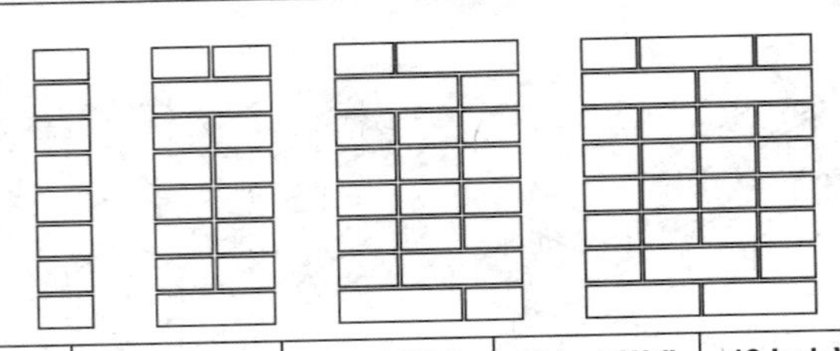

	4-Inch Wall		8-Inch Wall		12-Inch Wall		16-Inch Wall	
Area of Wall (sq. ft.)	No. of Brick	Cubic ft. of Mortar	No. of Brick	Cubic ft. of Mortar	No. of Brick	Cubic ft. of Mortar	No. of Brick	Cubic ft. of Mortar
1	6.2	0.075	12.4	0.195	18.5	0.314	24.7	0.433
10	62	1	124	2	185	3½	247	4½
20	124	2	247	4	370	6½	493	9
30	185	2½	370	6	555	9½	740	13
40	247	3½	493	8	740	13	986	17½
50	309	4	617	10	925	16	1,233	22
60	370	5	740	12	1,109	19	1,479	26
70	432	5½	863	14	1,294	22	1,725	31
80	493	6½	986	16	1,479	25	1,972	35
90	555	7	1,109	18	1,664	28	2,218	39
100	617	8	1,233	20	1,849	32	2,465	44
200	1,233	15	2,465	39	3,697	63	4,929	87
300	1,849	23	3,697	59	5,545	94	7,393	130
400	2,465	30	4,929	78	7,393	126	9,857	173
500	3,081	38	6,161	98	9,241	157	12,321	217

AMOUNT OF BRICK AND MORTAR NEEDED FOR VARIOUS WALL SIZES *(cont.)*

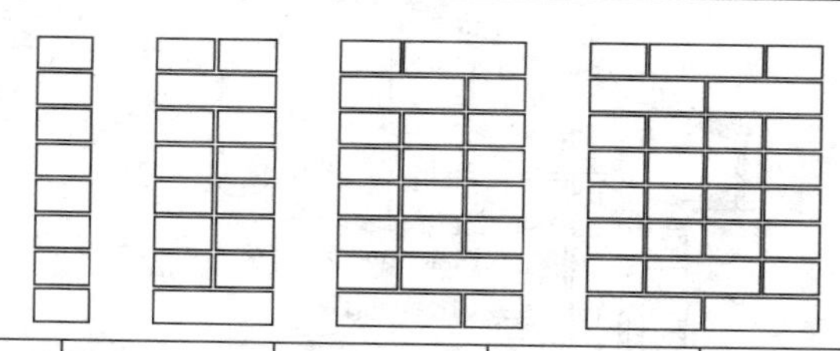

	4-Inch Wall		8-Inch Wall		12-Inch Wall		16-Inch Wall	
Area of Wall (sq. ft.)	**No. of Brick**	**Cubic ft. of Mortar**	**No. of Brick**	**Cubic ft. of Mortar**	**No. of Brick**	**Cubic ft. of Mortar**	**No. of Brick**	**Cubic ft. of Mortar**
600	3,697	46	7,393	117	11,089	189	14,786	260
700	4,313	53	8,625	137	12,937	220	17,250	303
800	4,929	61	9,857	156	14,786	251	19,714	347
900	5,545	68	11,089	175	16,634	283	22,178	390
1,000	6,161	76	12,321	195	18,482	314	24,642	433
2,000	12,321	151	24,642	390	36,963	628	49,284	866
3,000	18,482	227	36,963	584	55,444	942	73,926	1,299
4,000	24,642	302	49,284	779	73,926	1,255	98,567	1,732
5,000	30,803	377	61,605	973	92,407	1,568	123,209	2,165
6,000	36,963	453	73,926	1,168	110,888	1,883	147,851	2,599
7,000	43,124	528	86,247	1,363	129,370	2,197	172,493	3,032
8,000	49,284	604	98,567	1,557	147,851	2,511	197,124	3,465
9,000	55,444	679	110,888	1,752	166,332	2,825	221,776	3,898
10,000	61,605	755	123,209	1,947	184,813	3,139	246,418	4,331

Note: Mortar joints are ½ inch thick.

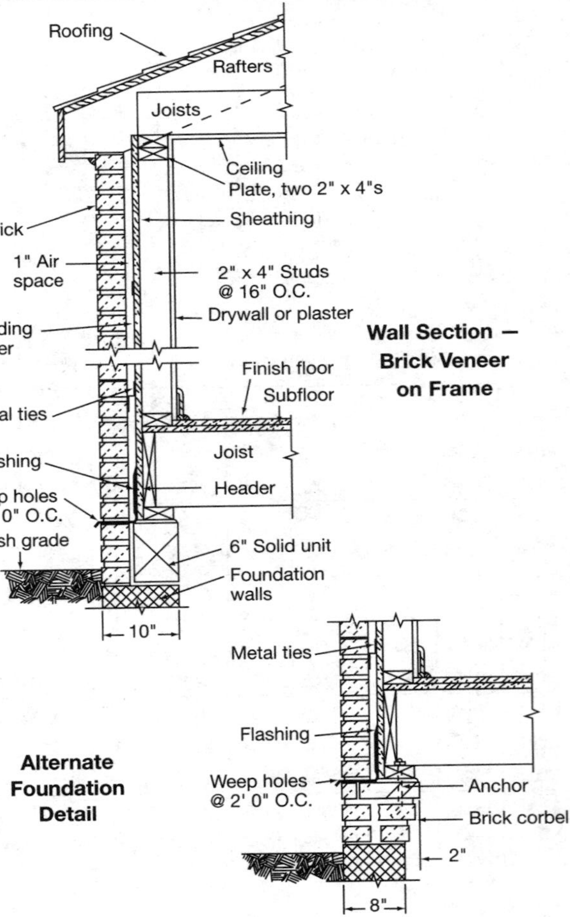

Roofing
Rafters
Joists
Ceiling
Plate, two 2" x 4"s
Brick
Sheathing
1" Air space
2" x 4" Studs @ 16" O.C.
Drywall or plaster
Building paper
Finish floor
Subfloor
Metal ties
Joist
Flashing
Header
Weep holes @ 2' 0" O.C.
Finish grade
6" Solid unit
Foundation walls
10"

Wall Section — Brick Veneer on Frame

Metal ties
Flashing
Weep holes @ 2' 0" O.C.
Anchor
Brick corbel
2"
8"

Alternate Foundation Detail

BRICK VENEER

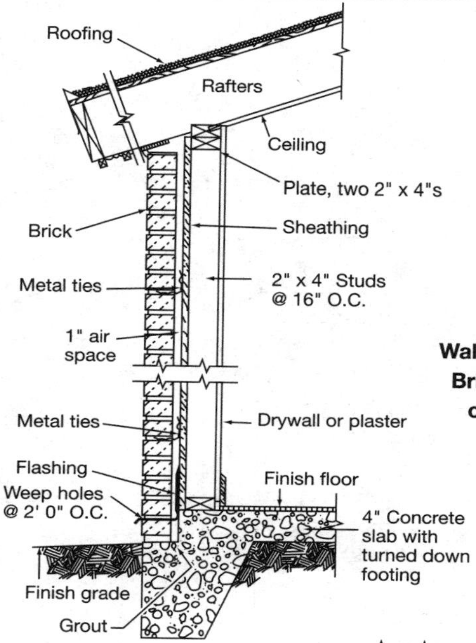

Wall Section — Brick Veneer on Frame

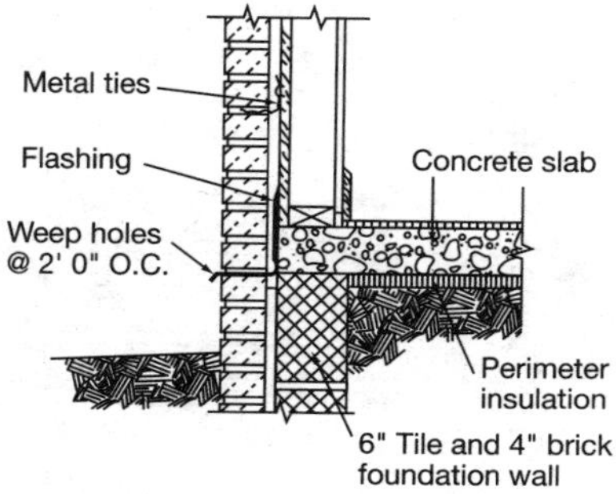

Alternate Foundation Detail

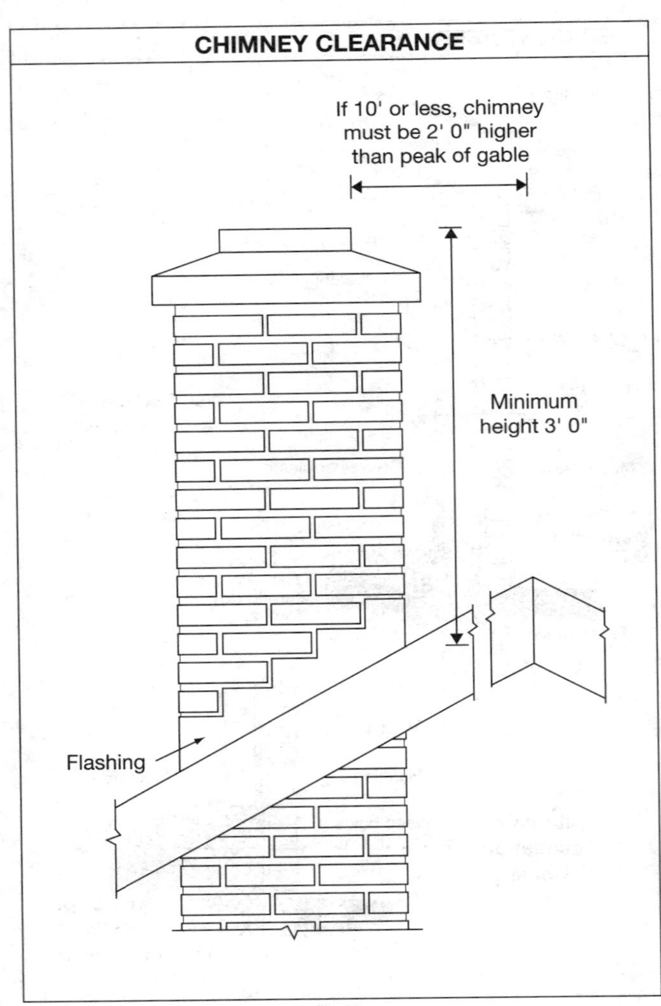
CHIMNEY CLEARANCE
If 10' or less, chimney
must be 2' 0" higher
than peak of gable
Minimum
height 3' 0"
Flashing

NUMBER OF BRICK FOR CHIMNEYS

per Foot of Height

Size and Number of Flues	Number of Brick	Cubic Feet Mortar
(a) 1-8" × 8" flue	27	0.5
(b) 1-8" × 12" flue	31	0.5
(c) 1-12" × 12" flue	35	0.6
(d) 2-8" × 8" flue	46	0.8
(e) 1-8" × 8" and 1-8" × 12" flue	51	0.9
(f) 2-8" × 12" flue	55	0.10
(g) 2-8" × 12" flue	53	0.9
(h) 1-8" × 12" and 12" × 12" flue	58	1.0
(i) 2-12" × 12" flue	62	1.1
(j) 2-8" × 8" and 1-8" × 12" flue	70	1.2
(k) 1-8" × 12" and 2-12" × 12" flue	83	1.4
(l) 1-8" × 12" and 2-12" × 12" flue	70	1.2
(m) 1-8" × 8" extending 12" from face of wall	18	0.4
(n) 1-8" × 8" extending 8" from face of wall	9	0.3
(o) 1-8" × 8" extending 4" from face of wall	0	0.0

SIZE OF FLUE LININGS			
Nominal Size of Chimney	Outside Dimensions of Flue Linings	Weight per Feet	Length of Piece
4" × 8"	4½" × 8½"	14 lbs.	2 feet
4" × 13"	4½" × 13"	20 lbs.	2 feet
8" × 8"	8½" × 8½"	18 lbs.	2 feet
8" × 12"	8½" × 13"	30 lbs.	2 feet
8" × 16"	8½" × 18"	36 lbs.	2 feet
12" × 12"	13" × 13"	38 lbs.	2 feet
12" × 16"	13" × 18"	45 lbs.	2 feet
16" × 16"	18" × 18"	62 lbs.	2 feet

CONCRETE MIXTURES BY VOLUME

Cement:Sand: Gravel Ratio	Application
1:3:6	Normal static loads, no rebar; not exposed
1:2.5:5	Normal foundations & walls; exposed
1:2.5:4	Basement walls
1:2.5:3.5	Waterproof basement walls
1:2.5:3	Floors, driveways
1:2.25:3	Steps, driveways, sidewalks
1:2:4	Lintels
1:2:4	Reinforced roads, buildings, walls; exposed
1:2:3.5	Retaining walls, driveways
1:2:3	Swimming pools, fence posts
1:1.75:4	Floors
1:1.5:3	Watertight, reinforced tanks & columns
1:1:2	High strength columns, girders, floors
1:1:1.5	Fence posts

Mix sand and cement first until a uniform color is obtained, then mix in the aggregate. Water volume will vary depending on the water content of the sand. Use only enough water to make the concrete mixture workable.

THICKNESS OF SLABS

Thickness (inches)	Application
4	Home basement floors, farm building floors
4 to 5	Home garage floors, porches
5 to 6	Sidewalks, barn floors, shed floors
6 to 8	Driveways

CUBIC VOLUMES OF SLABS

Cubic feet of Concrete = Slab thickness in feet × Slab width in feet × Slab length in feet

1 cubic yard = 27 cubic feet

1 cubic foot = 1,728 cubic inches

Note that the volume of the final concrete mixture is approximately 67% of the volume of the original cement–aggregate mixture. This is due to the sand and cement filling in the void spaces between the gravel fragments.

STEEL REINFORCING BAR

Bar Number	Diameter Fraction Inch	Diameter Inches	Diameter mm	Pounds per foot
2b	¼	0.250	6.4	0.17
3	⅜	0.375	9.5	0.38
4	½	0.500	12.7	0.67
5	⅝	0.625	15.9	1.04
6	¾	0.750	19.1	1.50
7	⅞	0.875	22.2	2.04
8	1	1.000	25.4	2.67
9	1–⅛	1.128	28.7	3.40
10	1–¼	1.270	32.3	4.30
11	1–⅜	1.410	35.8	5.31
14	1–¾	1.693	43.0	7.65
18	2–¼	2.257	57.3	13.60

CONCRETE AND MORTAR COLORS

Color	Color Material	lbs/Sack Cement
Black	Black oxide or mineral Black	1 to 12
Blue	Ultramarine Blue	5 to 9
Brown–Red	Red iron oxide	5 to 9
Bright Red	Mineral Turkey Red	5 to 9
Purple–Red	Indian red	5 to 9
Brown	Metallic Brown Oxide	5 to 9
Buff to yellow	Yellow ocher or yellow oxide	2 to 9
Green	Chromium oxide or ultramarine	5 to 9

CONCRETE SLABS, WALKS, AND DRIVEWAYS

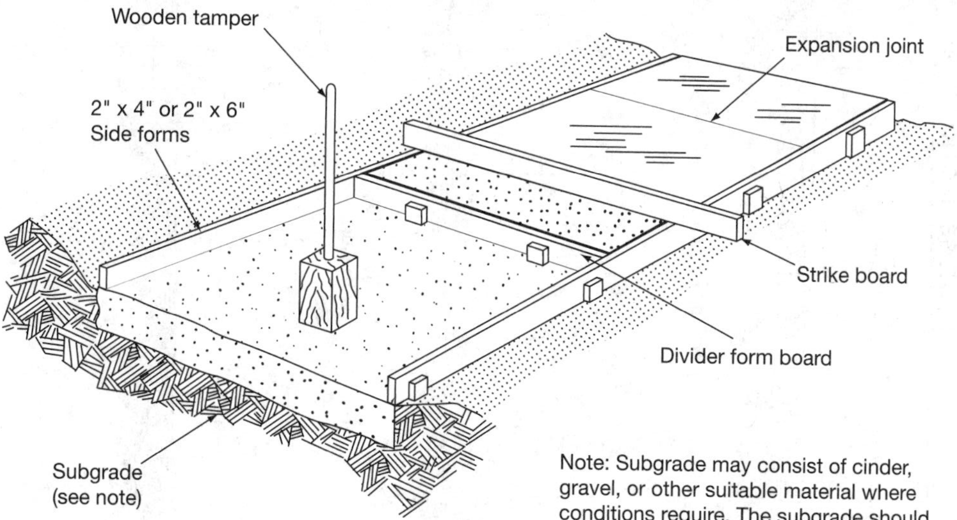

Wooden tamper

2" x 4" or 2" x 6"
Side forms

Expansion joint

Strike board

Divider form board

Subgrade
(see note)

Note: Subgrade may consist of cinder, gravel, or other suitable material where conditions require. The subgrade should be well-tamped before placing concrete.

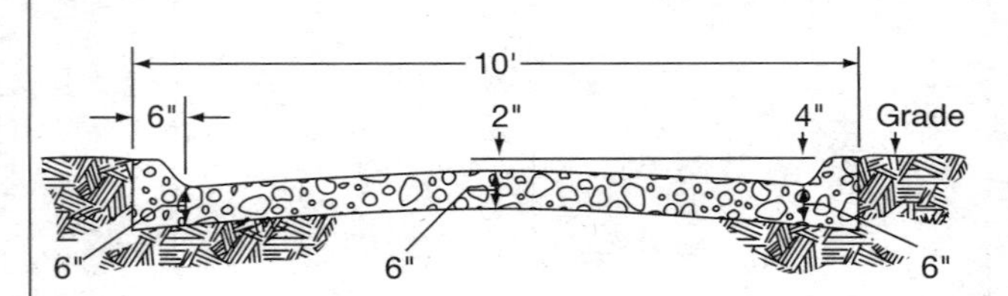

2" Crown

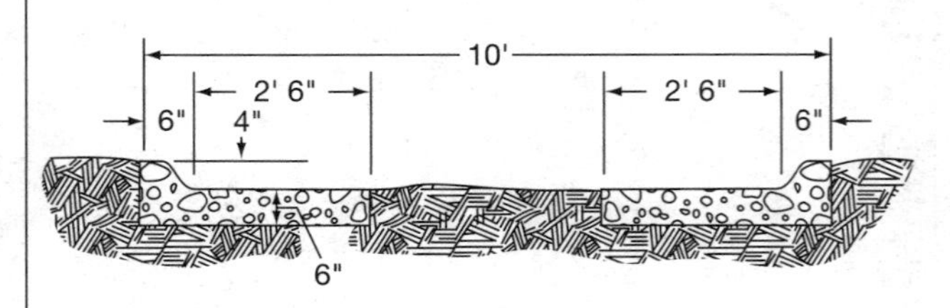

Parallel Strips
with Curb

Parallel Strips

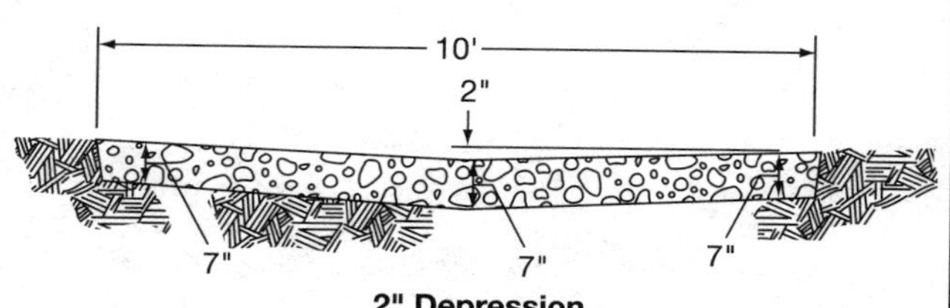

2" Depression

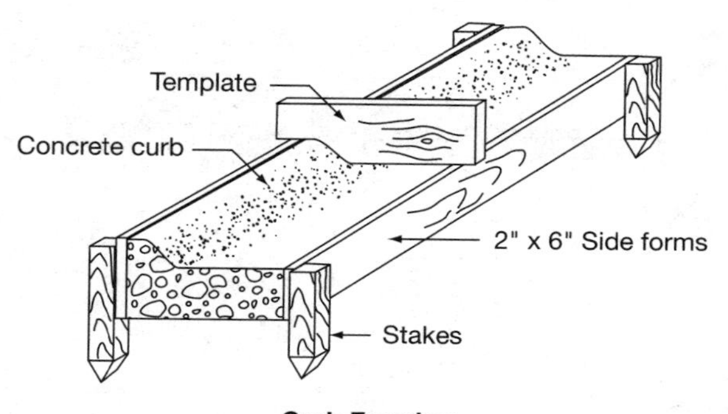

Curb Forming

CROSS-SECTION OF SINGLE-CAR LEVEL DRIVEWAY

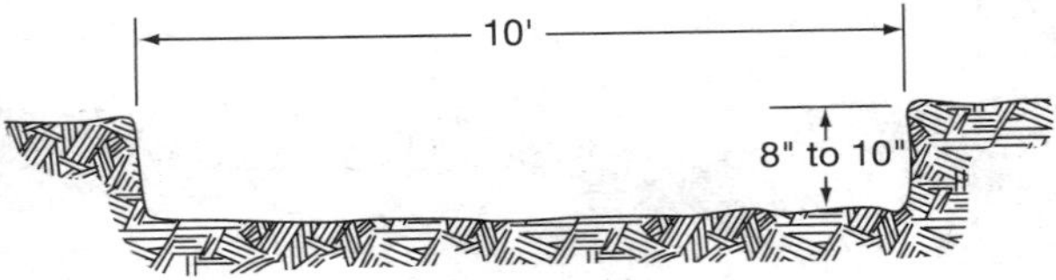

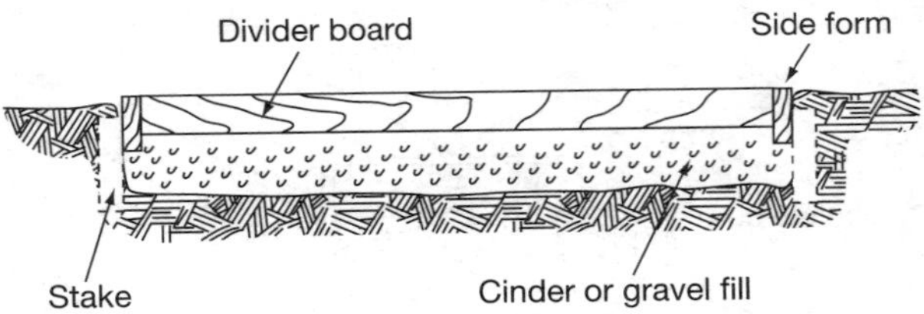

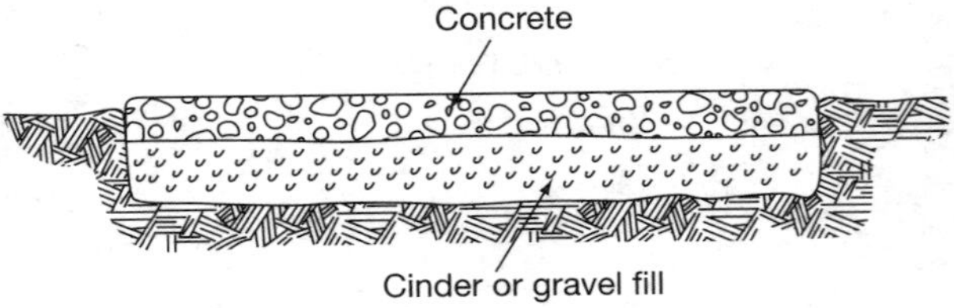

CHAPTER 2
Carpentry

LUMBER PRODUCT CLASSIFICATION		
	Thickness	**Width**
Board lumber	1"	2" or more
Light framing	2" to 4"	2" to 4"
Studs	2" to 4"	2" to 4" 10' and shorter
Structural light framing	2" to 4"	2" to 4"
Joists and planks	2" to 4"	6" and wider
Beams and stringers	5" and thicker	more than 2" greater than thickness
Posts and timbers	5" × 5" and larger	not more than 2" greater than thickness
Decking	2" to 4"	4" to 12" wide
Siding	thickness expressed by dimension of butt edge	
Moldings	size at thickest and widest points	
Note: Lengths of lumber generally are 6 feet and longer in multiples of 2 feet.		

SOFTWOOD LUMBER CLASSIFICATIONS AND GRADES

Boards

		Grade	IWP name
Appearance Grades	**Selects**	B & better C Select D Select	(IWP-Supreme) (IWP-Choice) (IWP-Quality)
Appearance Grades	**Finish**	Superior Prime E	
Appearance Grades	**Paneling**	Clear (any select or finish grade) No. 2 common selected for knotty paneling No. 3 common selected for knotty paneling	
Appearance Grades	**Siding (bevel, bungalow)**	Superior Prime	

	Grade	**Alternate Board Grades**
Boards Sheathing	No. 1 Common (IWP-Colonial) No. 2 Common (IWP-Sterling) No. 3 Common (IWP-Standard) No. 4 Common (IWP-Utility)	Select Merchantable Construction Standard Utility

Note: Names of grades and their specifications will vary.

Specification Check List

- ❏ Grades listed in order of quality.
- ❏ Include all species suited to project.
- ❏ For economy, specify lowest grade that will satisfy job requirement.
- ❏ Specify surface texture desired.
- ❏ Specify moisture content suited to project.
- ❏ Specify (WWP) grade stamp. For finish and exposed pieces, specify stamp on back or ends.

Western Red Cedar

Product	Grades
Finish Paneling and Ceiling	Clear Heart A B
Bevel Siding	Clear — V. G. Heart A — Bevel Siding B — Bevel Siding C — Bevel Siding

SOFTWOOD LUMBER CLASSIFICATIONS AND GRADES *(cont.)*			
Dimension			
Light Framing 2" to 4" thick 2" to 4" wide	Construction Standard Utility Economy	This category for use where high-strength values are NOT required; such as studs, plates, sills, cripples, blocking, etc.	
	Stud Economy Stud	An optional all-purpose grade limited to 10 feet and shorter. Characteristics affecting strength and stiffness values are limited so that the "Stud" grade is suitable for all stud uses, including load-bearing walls.	
Structural Light Framing 2" to 4" thick 2" to 4" wide	Select Structural No. 1 No. 2 No. 3 Economy	These grades are designed to fit those engineering applications where higher bending strength ratios are needed in light framing sizes. Typical uses would be for trusses, concrete pier wall forms, etc.	
Structural Joists and Planks 2" to 4" thick 6" and wider	Select structural No. 1 No. 2 No. 3 Economy	These grades are designed especially to fit in engineering applications for lumber 6 inches and wider, such as joists, rafters, and general framing uses.	
Timbers			
Beams and Stringers	Select Structural No. 1 No. 2 (No. 1 mining) No. 3 (No. 2 mining)	**Posts and Timbers**	Select Structural No. 1 No. 2 (No. 1 mining) No. 3 (No. 2 mining)

STANDARD LUMBER SIZES/NOMINAL, DRESSED

Product	Description	Nominal Size		Dressed Dimensions		
				Thicknesses and Widths (inches)		
		Thickness (inches)	Width (inches)	Surfaced Dry	Surfaced Unseasoned	Lengths (feet)
Framing	S4S	2 3 4	2 3 4 6 8 10 12 Over 12	1½ 2½ 3½ 5½ 7¼ 9¼ 11¼ Off ¾	1 9/16 2 9/16 3 9/16 5⅝ 7½ 9½ 11½ Off ½	6 ft. and longer in multiples of 1'
				Thickness (inches)	Width (inches)	
Timbers	Rough or S4S	5 and larger		½ off nominal		Same
		Nominal Size		Dressed Dimensions Surfaced Dry		
		Thickness (inches)	Width (inches)	Thickness (inches)	Width (inches)	Lengths (feet)
Decking Decking is usually surfaced to single T&G in 2" thickness and double T&G in 3" and 4" thicknesses	2" single T&G	2	6 8 10 12	1 ½	5 6¾ 8¾ 10¾	6 ft. and longer in multiples of 1'
	3" and 4" double T&G	3 4	6	2½ 3½	5¼	

Flooring	(D & M), (S2S & CM)	3/8 1/2 5/8 1 1 1/4 1 1/2	2 3 4 5 6	5/16 7/16 9/16 3/4 1 1 1/4	1 1/8 2 1/8 3 1/8 4 1/8 5 1/8	4 ft. and longer in multiples of 1'
Ceiling and Partition	(S2S & CM)	3/8 1/2 5/8 3/4	3 4 5 6	5/16 7/16 9/16 11/16	2 1/8 3 1/8 4 1/8 5 1/8	4 ft. and longer in multiples of 1'
Factory and Shop Lumber	S2S	1 (4/4) 1 1/4 (5/4) 1 1/2 (6/4) 1 3/4 (7/4) 2 (8/4) 2 1/2 (10/4) 3 (12/4) 4 (16/4)	5 and wider (4" and wider in 4/4 No. 1 Shop and 4/4 No. 2 Shop)	25/32 (4/4) 1 5/32 (5/4) 1 13/32 (6/4) 1 19/32 (7/4) 1 13/16 (8/4) 2 3/8 (10/4) 2 3/4 (12/4) 3 3/4 (16/4)	Usually sold random width	4 ft. and longer in multiples of 1'

Abbreviations

Abbreviated descriptions appearing in the size table are explained below.

S1S — Surfaced one side.
S2S — Surfaced two sides.
S4S — Surfaced four sides.
S1S1E — Surfaced one side, one edge.
S1S2E — Surfaced one side, two edges.
CM — Center matched.
D & M — Dressed and matched.
T & G — Tongue and grooved.
EV1S — Edge vee on one side.
S1E — Surfaced one edge.

HARDWOOD SIZES		
Nominal Size (fraction in)	**Rough Size (inches)**	**Surface 2 Sides Actual Size Dry (inches)**
4/4	1	13/16
5/4	1 1/4	1 1/16
6/4	1 1/2	1 5/16
7/4	1 3/4	1 1/2
8/4	2	1 3/4
10/4	2 1/2	2 1/4
12/4	3	2 3/4
14/4	3 1/2	3 1/4
16/4	4	3 3/4

HARDWOOD GRADES

First and Second (FAS) – Best grade. Required for a natural or stained finish. FAS boards must be at least 6 inches wide, 8 to 16 feet long, and 83.3% clear on the worst face.

Select – No. 1 Common – Minimum 3 inches wide, 4 to 16 feet long, 66.66% clear wood.

Select – No. 2 Common

Select – No. 3 Common

PLYWOOD IDENTIFICATION MARKINGS

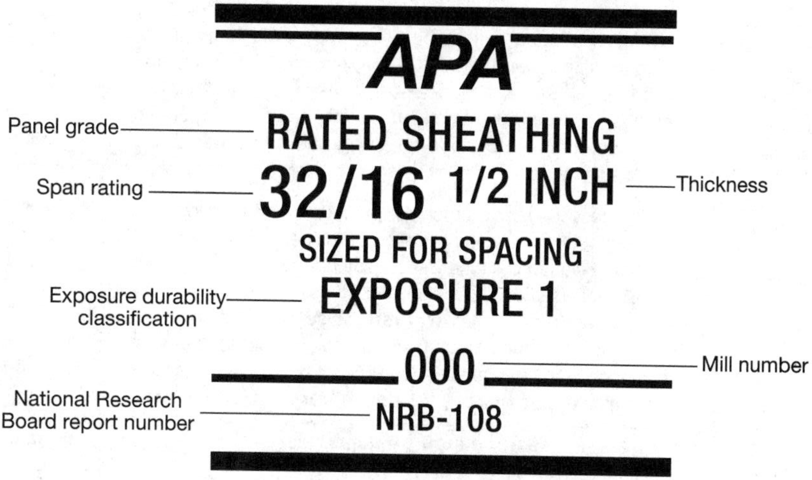

APA

Panel grade — RATED SHEATHING

Span rating — 32/16 1/2 INCH — Thickness

SIZED FOR SPACING

Exposure durability classification — EXPOSURE 1

000 — Mill number

National Research Board report number — NRB-108

Typical grade-trademark that is stamped on all plywood manufactured in compliance with panels national plywood standard.

PLYWOOD GRADES AND USAGE

Protected or Interior Use

Grade Designation	Description & Common Uses	Typical Trademarks
APA Rated Sheathing EXP 1 or 2	Specially designed for subflooring and wall and roof sheathing, but can also be used for a broad range of other construction and industrial applications. Can be manufactured as conventional veneered plywood, as a composite, or as a nonveneered panel. For special engineered applications, including high load requirements and certain industrial uses, veneered panels conforming to PS 1 may be required. Specify Exposure 1 when long construction delays are anticipated. Common thicknesses: 5/16", 3/8", 7/16", 1/2", 5/8", 3/4".	APA RATED SHEATHING 32/16 1/2 INCH SIZED FOR SPACING EXPOSURE 1 000 NRB-108
APA Structural I & II Rated Sheathing EXP 1	Unsanded all-veneer PS 1 plywood grades for use where strength properties are of maximum importance: structural diaphragms, box beams, gusset plates, stressed-skin panels, containers, pallet bins. Made only with exterior glue (Exposure 1). Structural I more commonly available. Common thicknesses: 5/16", 3/8", 1/2", 5/8", 3/4".	APA RATED SHEATHING STRUCTURAL I 24/0 3/8 INCH SIZED FOR SPACING EXPOSURE 1 000 PS 1-74 C-D INT/EXT GLUE NRB-108

APA Rated Sturd-I-Floor EXP 1 or 2	For combination subfloor-underlayment. Provides smooth surface for application of resilient floor covering and possesses high concentrated and impact load resistance. Can be manufactured as conventional veneered plywood, as a composite, or as a non-veneered panel. Available square edge or tongue-and-groove. Specify Exposure 1 when long construction delays are anticipated. Common thicknesses: 5/8" (19/32"), 3/4" (23/32").	APA RATED STURD-I-FLOOR 24/oc 23/32 INCH SIZED FOR SPACING T&G NET WIDTH 47-1/2 EXPOSURE 1 000 NRB-108
APA Rated Sturd-I-Floor 48 oc (2-4-1) EXP 1	For combination subfloor-underlayment on 32- and 48-inch spans and for heavy timber roof construction. Provides smooth surface for application of resilient floor coverings and possesses high concentrated and impact-load resistance. Manufactured only as conventional veneered plywood and only with exterior glue (Exposure 1). Available square edge or tongue-and-groove. Thickness: 1 1/8".	APA RATED STURD-I-FLOOR 48 oc 1-1/8 INCH (2-4-1) SIZED FOR SPACING EXPOSURE 1 T&G 000 INT/EXT GLUE NRB-108 FHA-UM-66
APA Rated Sheathing EXT	Exterior sheathing panel for subflooring and wall and roof sheathing, siding on service and farm buildings, crating, pallets, pallet bins, cable reels, etc. Manufactured as conventional veneered plywood. Common thicknesses: 5/16", 3/8", 1/2", 5/8", 3/4".	APA RATED SHEATHING 48/24 3/4 INCH SIZED FOR SPACING EXTERIOR 000 NRB-108

PLYWOOD GRADES AND USAGE *(cont.)*

Exterior Use

Grade Designation	Description & Common Uses	Typical Trademarks
APA Structural I & II Rated Sheathing EXT	For engineered applications in construction and industry where resistance to permanent exposure to weather or moisture is required. Manufactured only as conventional veneered PS I plywood. Unsanded. Structural I more commonly available. Common thicknesses: 5/16", 3/8", 1/2", 5/8", 3/4".	APA RATED SHEATHING STRUCTURAL I 42/20 5/8 INCH SIZED FOR SPACING EXTERIOR 000 PS 1-74 C-C NRB-108
APA Rated Sturd-I-Floor EXT	For combination subfloor-underlayment under resilient floor coverings where severe moisture conditions may be present, as in balcony decks. Possesses high concentrated and impact load resistance. Manufactured only as conventional veneered plywood. Available square edge or tongue-and-groove. Common thicknesses: 5/8" (19/32"), 3/4" (23/32").	APA RATED STURD-I-FLOOR 20 oc 19/32 INCH SIZED FOR SPACING EXTERIOR 000 NRB-108

Note: Specific grades, thicknesses, constructions, and exposure durability classifications may be in limited supply in some areas. Check with your supplier before specifying. Specify Performance-Rated Panels by thickness and Span Rating.

PLYWOOD VENEER GRADES	
N	Smooth surface "natural finish" veneer. Select, all heartwood or all sapwood. Free of open defects. Allows not more than 6 repairs, wood only, per 4 × 8 panel, made parallel to grain and well matched for grain and color.
A	Smooth, paintable. Not more than 18 neatly made repairs, boat, sled, or router type, and parallel to grain, permitted. May be used for natural finish in less demanding applications.
B	Solid surface. Shims, circular repair plugs, and tight knots to 1 inch across grain permitted. Some minor splits permitted.
C Plugged	Improved C veneer with splits limited to ⅛-inch width and knotholes and borer holes limited to ¼ × ½ inch. Admits some broken grain. Synthetic repairs permitted.
C	Tight knots to 1½ inch. Knotholes to 1 inch across grain and some to 1½ inch if total width of knots and knotholes is within specified limits. Synthetic or wood repairs. Discoloration and sanding defects that do not impair strength permitted. Limited splits allowed. Stitching permitted.
D	Knots and knotholes to 2½-inch width across grain and ½ inch larger within specified limits. Limited splits allowed. Stitching permitted. Limited to Interior, Exposure 1, and Exposure 2 panels.

NAIL SIZES

2d: 2 pennyweight × 1" long
4d: 4 pennyweight × 1¼" long
6d: 6 pennyweight × 1½" long
8d: 8 pennyweight × 1¾" long
10d: 10 pennyweight × 2" long

16d: 16 pennyweight × 3" long
20d: 20 pennyweight × 3½" long
straw nail = 10d x 6" long
spike = any nail 20d or larger × 6" long

SURFACE FINISHING ABBREVIATIONS, YARD AND STRUCTURAL LUMBER

Abbreviation	Definition
S1S	Smooth surface, one side
S1S1E	Smooth surface, one side, one end
S1S2E	Smooth surface, one side, two ends
S2S	Smooth surface, two sides
S2S1E	Smooth surface, two sides, one end
S2S2E	Smooth surface, two sides, two ends
S3S	Smooth surface, three sides
S3S1E	Smooth surface, three sides, one end
S3S2E	Smooth surface, three sides, two ends
S4S	Smooth surface, four sides
S4S1E	Smooth surface, four sides, one end
S4S2E	Smooth surface, four sides, two ends
R/E	Resawn (same as S1S or S1S2E)
R/O or R/S	Rough sawn (no smooth surfaces)

GLUE-LAMINATED TIMBER SIZING

Width		Depth	
Nominal	Actual	Nominal	Actual
3"	2¼"	8" (6¾")	9"
4"	3⅛"	10" (5⅛")	10½"
6"	5⅛"	8" (5⅛")	9"
8"	6¾"	6" (5⅛")	6"
10"	8¾"	6" (3⅛")	7½"
12"	10¾"		
14"	12¼"		
16"	14¼"		

DRYWALL SCREWS			
Description	**No.**	**Length (in.)**	**Applications**
Bugle Phillips	1E 2E 3E 4E 5R 6R 7R 8R	6 × 1 6 × 1⅛ 6 × 1¼ 6 × 1⅝ 6 × 2 6 × 2¼ 8 × 2½ 8 × 3	For attaching drywall to metal studs from 25 gauge through 20 gauge
Coarse Thread	1C 2C 3C 4C 5C 6C	6 × 1 6 × 1⅛ 6 × 1¼ 6 × 1⅝ 6 × 2 6 × 2¼	For attaching drywall to 25 gauge metal studs and attaching drywall to wood studs
Pan Framing	19	6 × 7/16	For attaching stud to track up to 20 gauge
HWH Framing	21 22 35	6 × 7/16 8 × 9/16 10 × ¾	For attaching stud to track up to 20 gauge where hex head is desired
K-Lath	28	8 × 9/16	For attaching wire lath K-lath to 20 gauge studs
Laminating	8	10 × 1½	Type G laminating screw for attaching gypsum to gypsum; a temporary fastener
Trim head	9 10	6 × 1⅝ 6 × 2¼	Trim head screw for attaching wood trim and base to 25 gauge studs

FASTENER APPLICATIONS

Nails — Common Applications

Joining	Size & Type	Placement
Wall Framing		
Top plate	8d common 16d common	
Header	8d common 16d common	
Header to joist	16d common	
Studs	8d common 16d common	
Wall Sheathing		
Boards	8d common	6" o.c.
Plywood (5/16", 3/8", 1/2")	6d common	6" o.c.
Plywood (5/8", 3/4")	8d common	6" o.c.
Fiberboard	1 3/4" galvanized roofing nail 8d galvanized common nail	6" o.c. 6" o.c.
Foamboard	Cap nail length sufficient for penetration of 1/2" into framing	12" o.c.
Gypsum	1 3/4" galvanized roofing nail 8d galvanized common nail	6" o.c. 6" o.c.
Subflooring	8d common	10"-12" o.c.
Underlayment	1 1/4" × 14 gauge annular underlayment nail	6" o.c. edges 12" o.c. face
Roof Framing		
Rafters, beveled or notched	12d common	
Rafter to joist	16d common	
Joist to rafter and stud	10d common	
Ridge beam	8d & 16d common	
Roof Sheathing		
Boards	8d common	
Plywood (5/16", 3/8", 1/2")	6d common	12" o.c. and 6" o.c. edges
Plywood (5/8", 3/4")	8d common	12" o.c. and 6" o.c. edges

FASTENER APPLICATIONS *(cont.)*		
Joining	**Size & Type**	**Placement**
Roofing, Asphalt		
New construction shingles and felt	⅞" through 1½" galvanized roofing	4 per shingle
Re-roofing application shingles and felt	1¾" or 2" galvanized roofing	4 per shingle
Roof deck/insulation	Thickness of insulation plus 1" insulation roof deck nail	
Roofing, Wood Shingles		
New construction	3d-4d galvanized shingle	2-3 per shingle
Re-roofing application	5d-6d galvanized shingle	2-3 per shingle
Soffit	6d-8d galvanized common	12" o.c. max.
Siding		
Bevel and lap Drop and shiplap Plywood	Aluminum nails are recommended for optimum performance	Consult siding manufacturer's application instructions
Hardboard	Galvanized hardboard siding nail Galvanized box nail	Consult siding manufacturer's application instructions
Doors, Windows, Moldings, Furring		
Wood strip to masonry Wood strip to stud or joist	Nail length is determined by thickness of siding and sheathing. Nails should penetrate at least 1½" into solid wood framing	
Paneling		
Wood	4d-8d casing-finishing	24" o.c.
Hardboard	2" × 16 gauge annular	8" o.c.
Plywood	3d casing-finishing	8" o.c.
Gypsum	1¼" annular drywall	6" o.c.
Lathing	4d common glued	4" o.c.
Exterior Projects		
Decks, patios, etc.	8d-16d hot dipped galvanized common	

NAILING SCHEDULE	
Connection	**Nailing**[1]
Joist to sill or girder, toenail	3-8d
Bridging to joist, toenail each end	2-8d
1" × 6" subfloor or less to each joist, face nail	2-8d
Wider than 1" × 6" subfloor to each joist, face nail	3-8d
2" subfloor to joist or girder, blind and face nail	2-16d
Sole plate to joist or blocking, face nail	16d at 16" o.c.
Top plate to stud, end nail	2-16d
Stud to sole plate	4-8d, toenail or 2-16d, end nail
Double studs, face nail	16d at 24" o.c.
Doubled top plates, face nail	16d at 16" o.c.
Top plates, laps and intersections, face nail	2-16d
Continuous header, two pieces	16d at 16" o.c. along each edge
Ceiling joists to plate, toenail	3-8d
Continuous header to stud, toenail	4-8d
Ceiling joists, laps over partitions, face nail	3-16d
Ceiling joists to parallel rafters, face nail	3-16d
Rafter to plate, toenail	3-8d
1" brace to each stud and plate, face nail	2-8d
1" × 8" sheathing or less to each bearing, face nail	2-8d
Wider than 1" × 8" sheathing to each bearing, face nail	3-8d
Built-up corner studs	16d at 24" o.c.
Built-up girder and beams	20d at 32" o.c. at top and bottom and staggered 2-20d at ends and at each splice
2" planks	2-16d at each bearing

NAILING SCHEDULE *(cont.)*

Connection	Nailing[1]
Plywood and Particleboard:[5]	
Subfloor, roof and wall sheathing (to framing):	
1/2" and less	6d[2]
19/32" – 3/4"	8d[3] or 6d[4]
7/8" – 1"	8d[2]
1 1/8" – 1 1/4"	10d[3] or 8d[4]
Combination Subfloor-underlayment (to framing):	
3/4" and less	6d[4]
7/8" – 1"	8d[4]
1 1/8" – 1 1/4"	10d[3] or 8d[4]
Panel Siding (to framing):	
1/2" or less	6d[6]
5/8"	8d[6]
Fiberboard Sheathing:[7]	
1/2"	No. 11 ga[8] 6d[3] No. 16 ga[9]
25/32"	No. 11 ga[8] 8d[3] No. 16 ga[9]

[1]Common or box nails may be used except where otherwise stated.

[2]Common or deformed shank.

[3]Common.

[4]Deformed shank.

[5]Nails spaced at 6 inches on center at edges, 12 inches at intermediate supports except 6 inches at all supports where spans are 48 inches or more. Nails for wall sheathing may be common, box, or casing.

[6]Corrosion-resistant siding or casing nails.

[7]Fasteners spaced 3 inches on center at exterior edges and 6 inches on center at intermediate supports.

[8]Corrosion-resistant roofing nails with 7/16-inch diameter head and 1 1/2-inch length for 1/2-inch sheathing and 1 3/4-inch length for 25/32-inch sheathing.

[9]Corrosion-resistant staples with nominal 7/16-inch crown and 1 1/8-inch length for 1/2-inch sheathing and 1 1/2-inch length for 25/32-inch sheathing.

SIZE, HEIGHT AND SPACING OF WOOD STUDS

Stud Size (inches)	Bearing Walls				Nonbearing Walls	
	Laterally Unsupported Stud Heights[1] (feet)	Supporting Roof and Ceiling Only	Supporting One Floor, Roof and Ceiling	Supporting Two Floors, Roof and Ceiling	Laterally Unsupported Stud Height[1] (feet)	Spacing (inches)
		Spacing (inches)				
2 × 3[2]	—	—	—	—	10	16
2 × 4	10	24	16	—	14	24
3 × 4	10	24	24	16	14	24
2 × 5	10	24	24	—	16	24
2 × 6	10	24	24	16	20	24

Utility grade studs shall be spaced no more than 16 inches on center, or support more than a roof and ceiling, or exceed 8 feet in height for exterior walls and load-bearing walls or 10 feet for interior nonbearing walls.

[1]Heights are distances between points of lateral support placed perpendicular to the plane of the wall. Increases in unsupported height are permitted where justified by an analysis.

[2]Shall not be used in exterior walls.

NUMBER AND SPACING OF WOOD JOISTS FOR ANY FLOOR

Length of Span	Spacing of Joists									
	12"	16"	20"	24"	30"	36"	42"	48"	54"	60"
6	7	6	5	4	3	3	3	3	2	2
7	8	6	5	5	4	4	3	3	3	2
8	9	7	6	5	4	4	3	3	3	3
9	10	8	6	6	5	4	4	3	3	3
10	11	9	7	6	5	4	4	4	3	3
11	12	9	8	7	5	5	4	4	3	3
12	13	10	8	7	6	5	4	4	4	3
13	14	11	9	8	6	5	5	4	4	4
14	15	12	9	8	7	6	5	5	4	4
15	16	12	10	9	7	6	5	5	4	4
16	17	13	11	9	7	6	6	5	5	4
17	18	14	11	10	8	7	6	5	5	4
18	19	15	12	10	8	7	6	6	5	4
19	20	15	12	11	9	7	6	6	5	5
20	21	16	13	11	9	8	7	6	5	5
21	22	17	14	12	9	8	7	6	6	5
22	23	18	14	12	10	8	7	7	6	5
23	24	18	15	13	10	9	8	7	6	6
24	25	19	15	13	11	9	8	7	6	6
25	26	20	16	14	11	9	8	7	7	6
26	27	21	17	14	11	10	8	8	7	6
27	28	21	17	15	12	10	9	8	7	6
28	29	22	18	15	12	10	9	8	7	7
29	30	23	18	16	13	11	9	8	7	7
30	31	24	19	16	13	11	10	9	8	7
31	32	24	20	17	13	11	10	9	8	7
32	33	25	20	17	14	12	10	9	8	7
33	34	26	21	18	14	12	10	9	8	8
34	35	27	21	18	15	12	11	10	9	8
35	36	27	22	19	15	13	11	10	9	8
36	37	28	23	19	15	13	11	10	9	8
37	38	29	23	20	16	13	12	10	9	8
38	39	30	24	20	16	14	12	11	9	9
39	40	30	24	21	17	14	12	11	10	9
40	41	31	25	21	17	14	12	11	10	9

One joist has been added to each of the above quantities to take care of extra joist required at end of span. Add for doubling joists under all partitions.

BUILT-UP WOOD HEADER DOUBLE OR TRIPLE ON TWO 4" × 4" POSTS								
Span (in feet)	Weight (in pounds) Safely Supported by:							
	2 - 2 × 6	2 - 2 × 8	2 - 2 × 10	2 - 2 × 12	3 - 2 × 6	3 - 2 × 8	3 - 2 × 10	3 - 2 × 12
4	2,250	4,688	5,000	5,980	3,780	5,850	7,410	8,970
6	1,680	3,126	5,000	5,980	2,520	4,689	7,410	8,970
8	—	2,657	3,761	5,511	—	3,985	5,641	8,266
10	—	2,125	3,008	4,409	—	3,187	4,512	6,613
12	—	—	2,507	3,674	—	—	3,760	5,511
14	—	—	—	3,149	—	—	—	4,723

ALLOWABLE SPANS FOR HEADERS UNDER DIFFERENT LOAD CONDITIONS

	Outside Walls			Inside Walls			
Nominal Depth of Header (inches)	Roof with or without Attic Storage	Roof with or without Attic Storage + One Floor	Roof with or without Attic Storage + Two Floors	Little Attic Storage	Full Attic Storage, or Roof Load, or Little Attic Storage + One Floor	Full Attic Storage + One Floor, or Roof Load + One Floor, or Little Attic Storage + Two Floors	Full Attic Storage + Two Floors, or Roof Load + Two Floors
4	4'	2'	2'	4'	2'	No	No
6	6'	5'	4'	6'	3'	2' 6"	2'
8	8'	7'	6'	8'	4'	3'	3'
10	10'	8'	7'	10'	5'	4'	3' 6"
12	12'	9'	8'	12' 6"	6'	5'	4'

STEEL PLATE HEADER ON TWO 4" × 4" POSTS									
Plate / Span (in feet)	Weight (in pounds) Safely Supported by Wood Sides and Plate								
	2 – 2 × 8 + 7½" by			2 – 2 × 10 + 9½" by			2 – 2 × 12 + 11½" by		
	3/8"	7/16"	1/2"	3/8"	7/16"	1/2"	3/8"	7/16"	1/2"
10	6,754	7,538	8,242	10,973	12,199	13,418	15,933	17,729	19,604
12	5,585	6,216	6,827	9,095	10,131	11,106	13,224	14,517	16,265
14	4,756	5,293	5,811	7,751	8,623	9,463	11,295	12,561	13,876
16	–	4,481	5,036	6,746	7,494	8,221	9,815	10,953	12,086
18	–	–	–	5,942	6,606	7,158	8,675	9,652	10,647
20	–	–	–	–	–	6,466	7,746	8,618	9,408

STANDARD DOOR OPENING SIZES							
Width of Opening	1¾-Inch Thick Doors Height of Opening					1⅜-Inch Thick Doors Height of Opening	
2' 0"	6' 8"	7' 0"	7' 2"	7' 10"	8' 0"	6' 8"	7' 0"
2' 4"	6' 8"	7' 0"	7' 2"	7' 10"	8' 0"	6' 8"	7' 0"
2' 6"	6' 8"	7' 0"	7' 2"	7' 10"	8' 0"	6' 8"	7' 0"
2' 8"	6' 8"	7' 0"	7' 2"	7' 10"	8' 0"	6' 8"	7' 0"
3' 0"	6' 8"	7' 0"	7' 2"	7' 10"	8' 0"	6' 8"	7' 0"
3' 4"	6' 8"	7' 0"	7' 2"	7' 10"	8' 0"	—	—
3' 6"	6' 8"	7' 0"	7' 2"	7' 10"	8' 0"	—	—
3' 8"	6' 8"	7' 0"	7' 2"	7' 10"	8' 0"	—	—
4' 0"	6' 8"	7' 0"	7' 2"	7' 10"	8' 0"	—	—

DOOR AND FRAME SELECTOR—TYPE OF DOOR FRAME										
Door Weight	Gauge Jamb Studs	Bracing Over Header	Knock Down Alum.	Knock Down Steel	Fixed Steel	Requires Mechanical Closure	Height[a] Stud Size			
							1⅝"	2½"	3⅝"	4"
Up to 50 lbs.	25		x				NA[b]	10'	14'	15'
	25			x			NA	10'	14'	15'
	25				x		NA	10'	14'	15'
	25	x	x				10'	12'	16'	17'
	25	x		x			10'	12'	16'	17'
	25	x			x		10'	12'	16'	17'
	20	x	x				—	16'	21'	22'
	20	x		x			—	16'	21'	22
	20	x			x		—	16'	21'	22'

	25	x			x	x	12'	10'	14'	15'
	25	x			x			10'	14'	15'
	25	x		x				10'	14'	15'
	25	x	x			x	10'	12'	14'	15'
50 lbs.	20		x			x		10'	14'	15'
to	20			x				10'	14'	15'
80 lbs.	20				x			10'	14'	15'
	20	x		x				12'	16'	17'
	20	x			x			12'	16'	17'
	20 dbl.	x		x				16'	21'	22'
	20 dbl.				x			16'	21'	22'
80 lbs.	25	x			x			12'	16'	17'
to	20	x	x			x		12'	16'	17'
120 lbs.	20			x		x		12'	16'	17'
	20 dbl.	x			x			16'	21'	22'

[a]See partition height table for stud spacing.
[b]NA-Not allowed.

EXTERIOR DOOR DETAILS

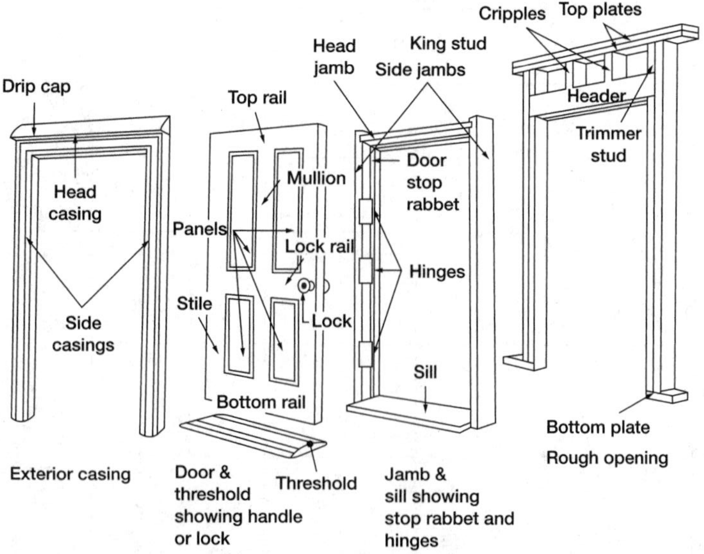

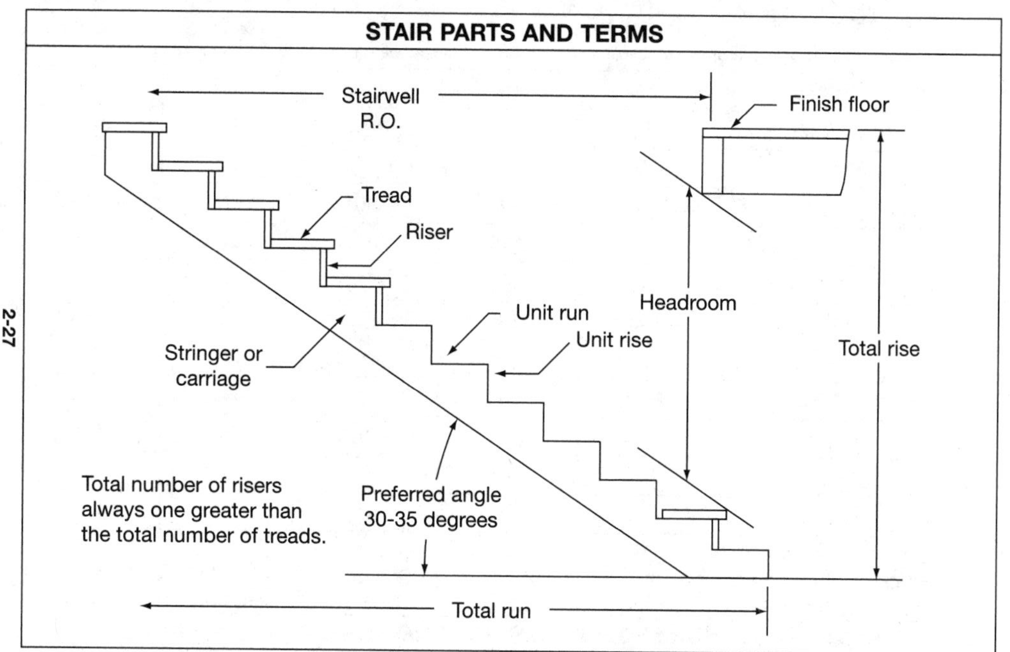
STAIR PARTS AND TERMS
Stairwell
R.O.
Finish floor
Tread
Riser
Headroom
Unit run
Unit rise
Total rise
Stringer or
carriage
Total number of risers
always one greater than
the total number of treads.
Preferred angle
30-35 degrees
Total run

DIMENSIONS FOR STRAIGHT STAIRS

Height Floor-to-Floor H	Number of Risers	Height of Risers R	Width of Treads T	Total Run L	Minimum Headroom Y	Well Opening U
8' 0"	12	8"	9"	8' 3"	6' 6"	8' 1"
8' 0"	13	$7^3/_8$" +	$9^1/_2$"	9' 6"	6' 6"	9' $2^1/_2$"
8' 0"	13	$7^3/_8$" +	10"	10' 0"	6' 6"	9' $8^1/_2$"
8' 6"	13	$7^7/_8$" –	9"	9' 0"	6' 6"	8' 3"
8' 6"	14	$7^5/_{16}$" –	$9^1/_2$"	10' $3^1/_2$"	6' 6"	9' 4"
8' 6"	14	$7^5/_{16}$" –	10"	10' 10"	6' 6"	9' 10"
9' 0"	14	$7^{11}/_{16}$" +	9"	9' 9"	6' 6"	8' 5"
9' 0"	15	$7^3/_{16}$" +	$9^1/_2$"	11' 1"	6' 6"	9' $6^1/_2$"
9' 0"	15	$7^3/_{16}$" +	10"	11' 8"	6' 6"	9' $11^1/_2$"
9' 6"	15	$7^5/_8$" –	9"	10' 6"	6' 6"	8' $6^1/_2$"
9' 6"	16	$7^1/_8$"	$9^1/_2$"	11' $10^1/_2$"	6' 6"	9' 7"
9' 6"	16	$7^1/_8$"	10"	12' 6"	6' 6"	10' 1"

Dimensions shown under well opening "U" are based on 6-foot, 6-inch minimum headroom. If headroom is increased, well opening also increases.

STAIR RISERS, TREADS AND DIMENSIONS

Total Rise Floor-to-Floor H	Number of Risers	Height of Riser R	Number of Treads	Width of Run T	Total Run L	Well Opening U	Length of Carriage	Use Stock Tread Width	Dimension of Nosing Projection
8' 0"	12	8"	11	9 1/2"	8' 8 1/2"	9' 1"	11' 4 5/8"	10 1/2"	1"
	14	6 7/8"	13	10 5/8"	11' 6 1/8"	10' 10"	13' 8 1/2"	11 1/2"	7/8"
8' 4"	13	7 11/16"	12	9 13/16"	9' 9 3/4"	10' 0"	12' 5 1/2"	10 1/2"	11/16"
	14	7 1/8"	13	10 3/8"	11' 2 7/8"	11' 0"	13' 7 5/8"	11 1/2"	1 1/8"
8' 6"	13	7 7/8"	12	9 5/8"	9' 7 1/2"	9' 2"	12' 5 1/4"	10 1/2"	7/8"
	14	7 5/16"	13	10 3/16"	11' 1/2"	10' 8"	13' 7"	11 1/2"	1 5/16"
8' 9"	14	7 1/2"	13	9 1/4"	10' 1/4"	9' 5"	12' 10 3/4"	10 1/2"	1 1/4"
	14	7 1/2"	13	10"	10' 10"	10' 1"	13' 6 1/2"	11 1/2"	1 1/2"
8' 11"	14	7 5/8"	13	9 3/8"	10' 1 7/8"	9' 5"	13' 1 1/4"	10 1/2"	1 1/8"
	14	7 5/8"	13	9 1/16"	9' 9 7/8"	9' 0"	12' 10"	10 1/2"	1 7/16"
	14	7 5/8"	13	10 1/4"	11' 1 1/4"	10' 2"	13' 10 1/4"	11 1/2"	1 1/4"
9' 1"	14	7 13/16"	13	9 11/16"	10' 6"	9' 5"	13' 5 3/4"	10 1/2"	13/16"
	15	7 1/4"	14	10 1/4"	11' 11 1/2"	10' 8"	14' 7 3/4"	11 1/2"	1 1/4"

Well openings based on minimum head height of 6' 8". Dimensions based on 2" × 10" floor joist.

TREATED STRESSED SKIN PANEL CONSTRUCTION

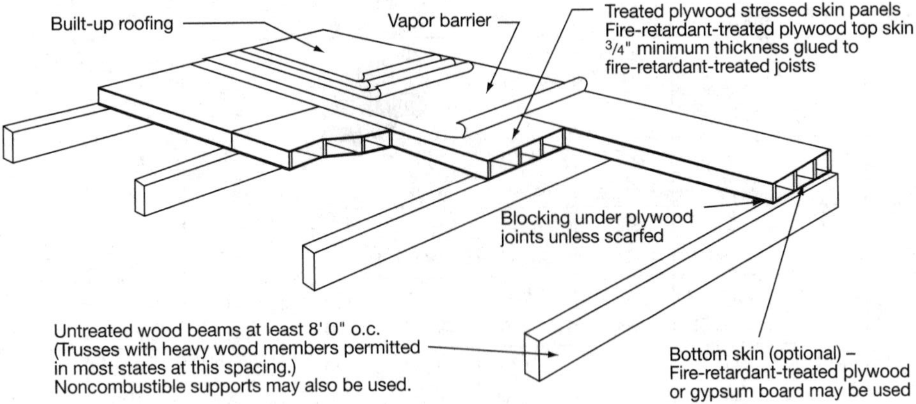

Built-up roofing

Vapor barrier

Treated plywood stressed skin panels
Fire-retardant-treated plywood top skin
3/4" minimum thickness glued to
fire-retardant-treated joists

Blocking under plywood
joints unless scarfed

Untreated wood beams at least 8' 0" o.c.
(Trusses with heavy wood members permitted
in most states at this spacing.)
Noncombustible supports may also be used.

Bottom skin (optional) –
Fire-retardant-treated plywood
or gypsum board may be used

Notes:
1. Aluminum foil vapor barrier required only for NM 501 construction.
2. For NM 501 construction, use tongue-and-groove plywood joints or treated blocking.

ONE-HOUR INTERIOR SHEAR WALL CONSTRUCTION

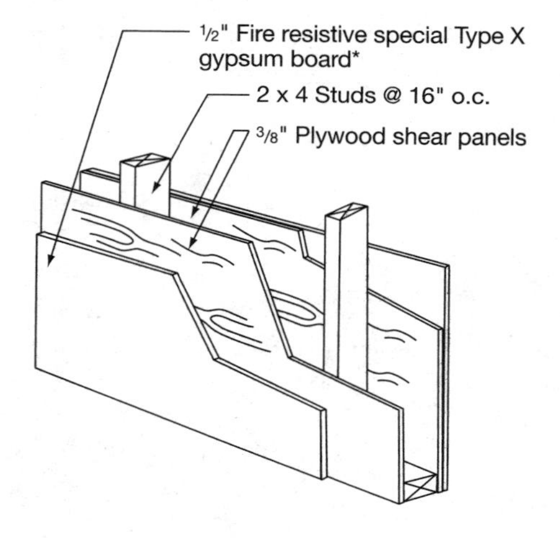

*Regular 1/2" gypsum board may be used when mineral wool or glass fiber batts are used in wall cavity.

Insulation batts in wall cavity also used for sound transmission control.

ONE-HOUR ASSEMBLY T-BAR GRID CEILING SYSTEM

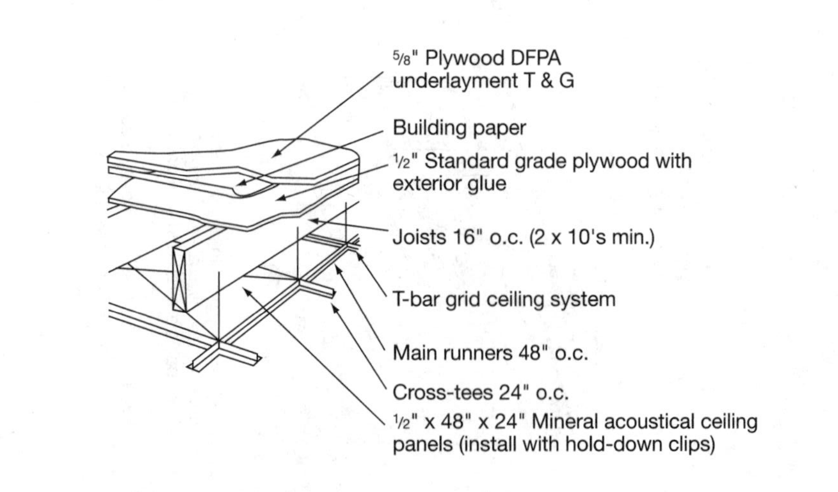

ONE-HOUR ASSEMBLY RESILIENT CHANNEL CEILING SYSTEM

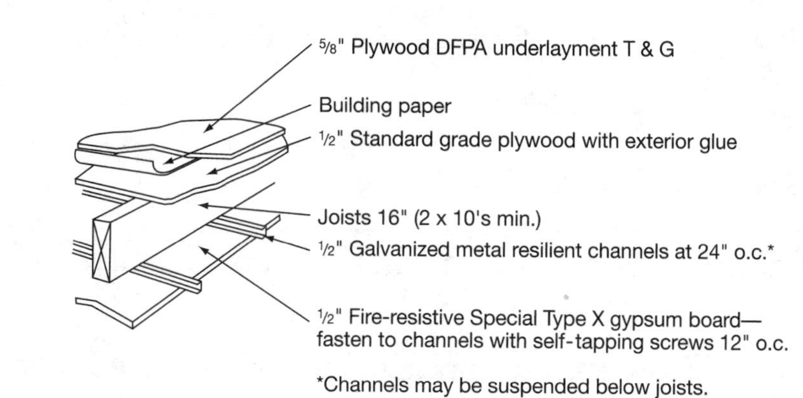

*Channels may be suspended below joists.

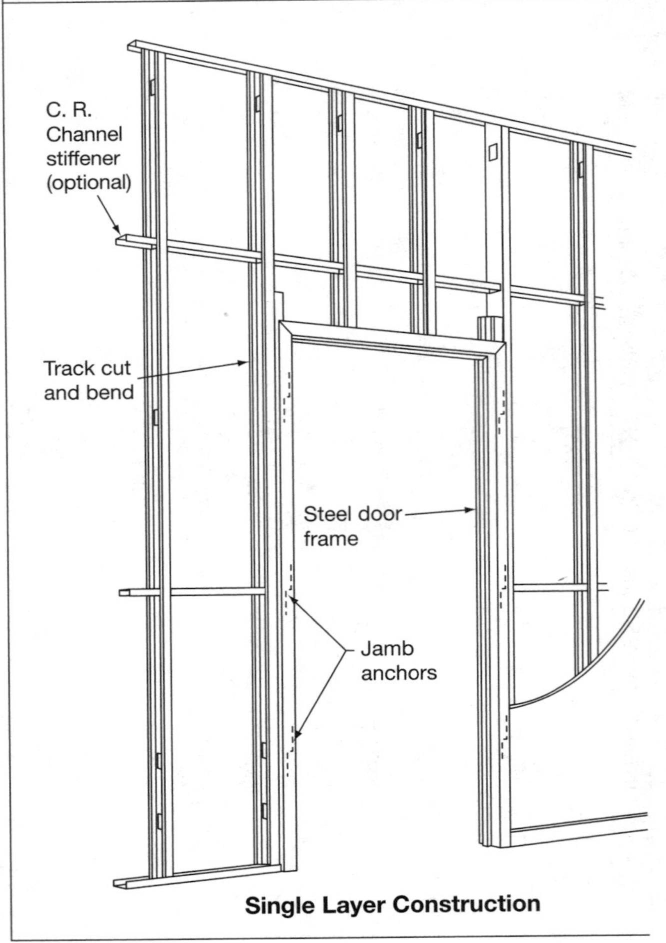

Single Layer Construction

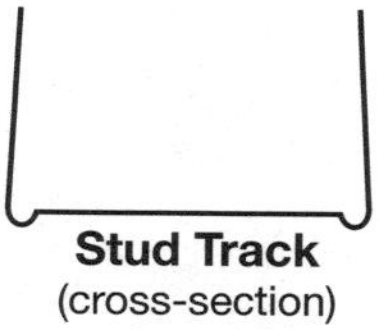

Stud Track
(cross-section)

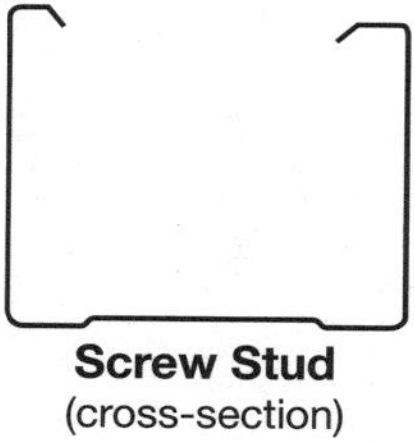

Screw Stud
(cross-section)

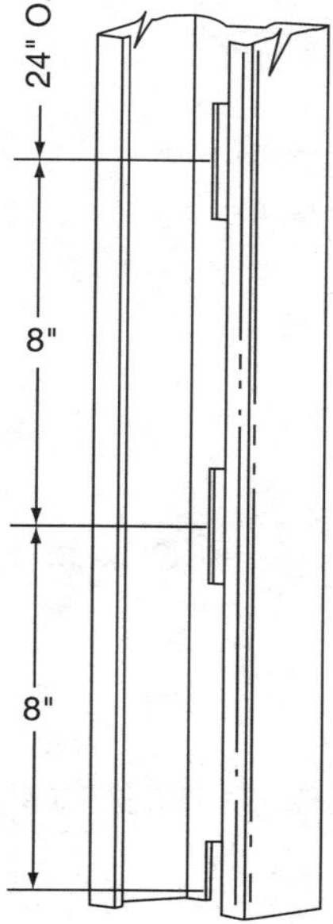

24" O.C. (nominal) continuously

8"

8"

Screw Stud

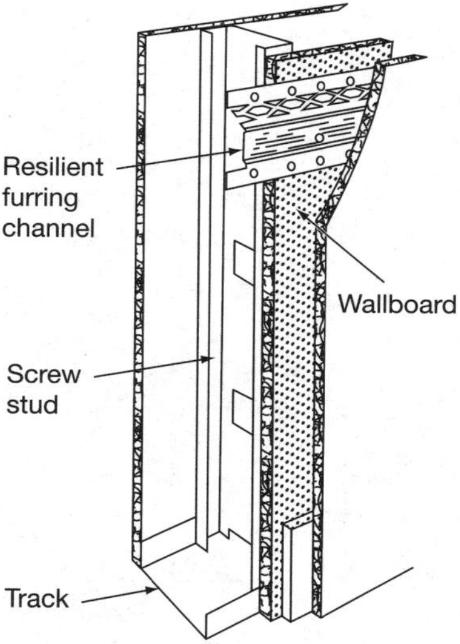

Resilient furring channel

Wallboard

Screw stud

Track

Resilient Furring Channel with Gypsum Wallboard

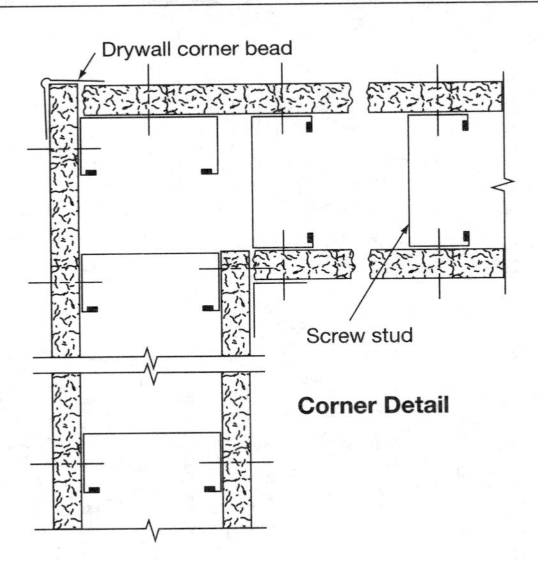

Corner Detail

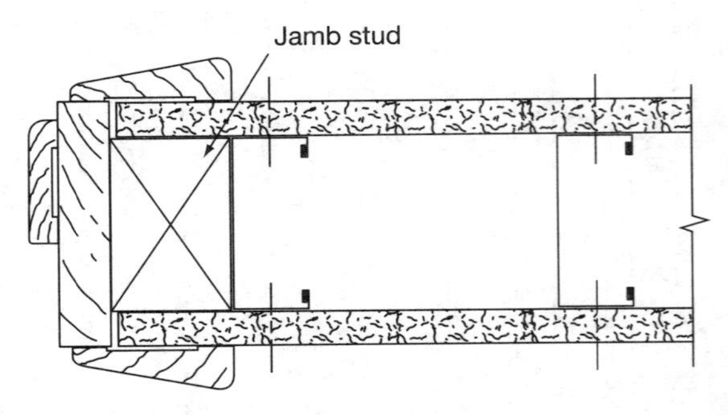

Wood Door Jamb Detail

JAMBS—METAL WALL CONSTRUCTION

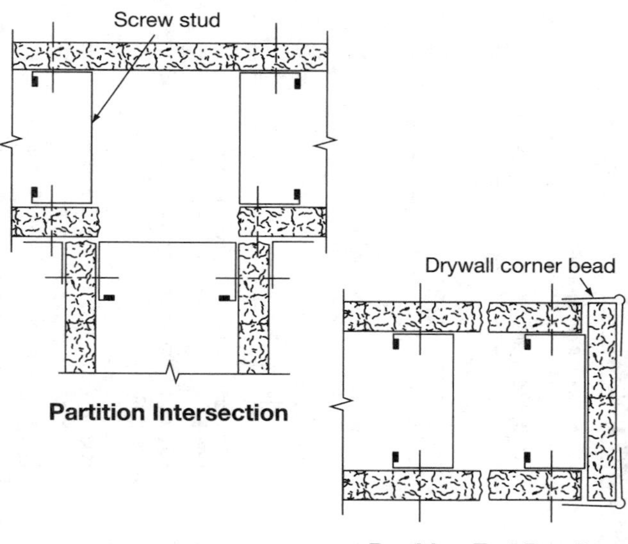

Partition Intersection

Partition End Detail

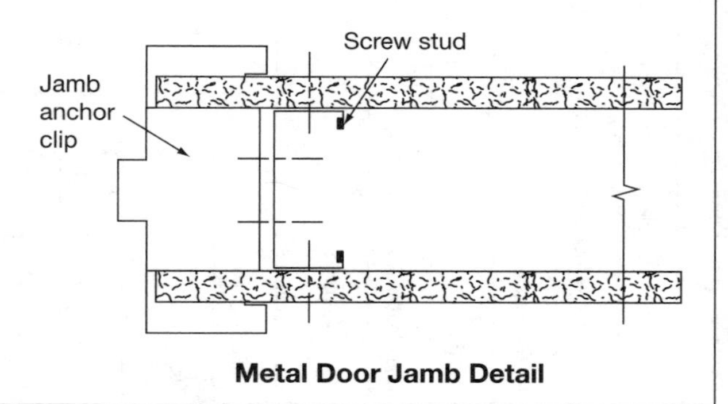

Metal Door Jamb Detail

HARDWARE LOCATIONS	
Locks, Latches, Roller Latches, and Double Handle Sets	Centerline of lock strike 40 5/16" from bottom of frame
Cylindrical or Mortise Deadlocks	Centerline of strike 60" from bottom of frame
Push Plates	Centerline 45" from bottom of frame
Pull Plates	Centerline of grip 42" from bottom of frame
Combination Push Bar	Centerline 42" from bottom of frame
Hospital Arm Pull	Centerline of lower base is 45" from bottom of frame with grip open at bottom
Panic Devices	Centerline of strike 40 5/16" from bottom of frame
Top Hinge	Up to 11 3/4" from rabbet section of head of frame to centerline of hinge
Bottom Hinge	Up to 13" from bottom of frame to centerline of hinge
Intermediate Hinge(s)	Equally spaced between top and bottom hinge

HARDWARE REINFORCING GAUGES

Hardware			Minimum Gauge
Hinges	1¾"	Door	10 gauge or equivalent number of threads 12 gauge allowed if reinforcing is channel shaped
	1¾"	Frame	10 gauge or equivalent number of threads
	1⅜"	Door	12 gauge or equivalent number of threads
	1⅜"	Frame	12 gauge or equivalent number of threads
Mortise Locksets and Deadlocks		Door	14 gauge or equivalent number of threads
		Frame	14 gauge or equivalent number of threads
Bored or Cylindrical Locks		Door	14 gauge or equivalent number of threads
		Frame	14 gauge or equivalent number of threads
Flush Bolts and Chain and Foot Bolts		Door	14 gauge
		Frame	14 gauge
Surface-Applied Closers		Door	14 gauge
		Frame	14 gauge
Hold-Open Arms		Door	14 gauge
		Frame	14 gauge
Pull Plates and Bars		Door	16 gauge except when through bolts are used
Kick and Push Plates and Bars		Door	Not required
Surface Panic Devices		Door	14 gauge
		Frame	14 gauge
Floor Checking Hinges and Pivots		Door	7 gauge
		Frame	7 gauge

STANDARD STEEL DOOR GRADES AND MODELS

Grade	Model	Construction	Minimum Thickness Gauge Number Panels – Face Sheet	Minimum Thickness Gauge Number Stiles – Rails
I – Standard Duty (1¾" and 1⅜")	1	Full Flush (Hollow Steel)	20	–
	2	Full Flush (Composite)	20	–
	3	Seamless (Hollow Steel)	20	–
	4	Seamless (Composite)	20	–
II – Heavy Duty (1¾" only)	1	Full Flush (Hollow Steel)	18	–
	2	Full Flush (Composite)	18	–
	3	Seamless (Hollow Steel)	18	–
	4	Seamless (Composite)	18	–
III – Extra Heavy Duty (1¾" only)	1	Full Flush (Hollow Steel)	16	–
	2	Full Flush (Composite)	16	–
	3	Seamless (Hollow Steel)	16	–
	4	Seamless (Composite)	16	–
	5	Flush Panel (Stile and Rail)	18	16

FIRE DOOR CLASSIFIED OPENINGS

Opening	Class	Rating		Glass
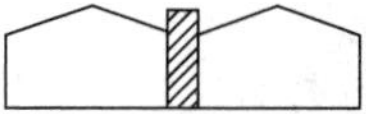	A	3 Hours	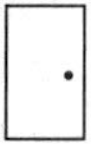	None
	B	1½ Hours		100 sq. in. per door leaf
	C	¾ Hour	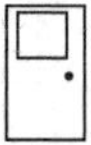	1,296 sq. in. per light
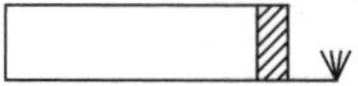	D	1½ Hours	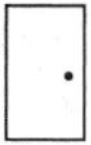	None
	E	¾ Hour	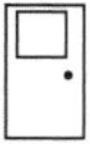	1,296 sq. in. per light
	No class designation	20 Minutes	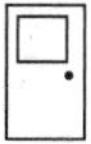	1,296 sq. in. per light

REPLACEMENT WINDOWS		
Material	**Finish and Durability**	**Preferences**
Aluminum Silicon-magnesium-aluminum wrought alloy offers corrosion resistance. Roll-formed alloys are not as durable.	Baked synthetic enamel is best. Fluorocarbon-coated enamel is second best. Good: Anodized or paint finish is good. Oxidized mill finish is not.	¼ inch wide wood or plastic thermal break; tight weather stripping; polyvinylchloride or neoprene gaskets; aluminum, plastic, or stainless steel fittings.
Steel	Needs baked-on synthetic enamel as a base metal. Conducts heat and is subject to condensation. Requires annual painting for long-term durability. Marble or laminate windowsills are recommended to avoid moisture damage.	Complete paint coverage, no scratches. Single structure without thermal break.
Wood	Painted windows require maintenance, vinyl-clad do not.	Low-E glass, ¼ inch or more spacing between panes. Tight seals.

WINDOWS AND DOOR INSTALLATION TOOLS			
	Hand Tools		**Power Tools (optional)**
Installing Door and Lock	Awl Brace and bits Caulking gun Chisel Compass Gauge Hammer	Ladder Pencil Plane Rafter square Screwdriver Spirit level Steel rule	Circular or arbor saw Drill Router Hole saw
Installing Window	Awl Caulking gun Chisel Drill Gauge Hammer	Ladder Screwdriver Steel rule Spirit level	Drill
Installing or Replacing Windowpane	Chisel Glass Glass cutter Glazier's point Hammer	Putty knife Rafter square Sandpaper Soldering iron Straightedge	
Replacing Sash Cord	Chain or cord Chisel Hammer Knife Screwdrivers Screws String Weight		
Loosening Binding, Window or Door	Chisel File Hammer Plane Pry bar Putty knife Screwdriver	Sander Petroleum wax or silicone	

CLASSIFICATIONS OF FLOAT GLASS (LISTED IN DESCENDING QUALITY)		
Class 1–Transparent	**Class 2–Heat-Absorbing & Light-Reducing**	**Class 3–Light-Reducing, Tinted**
q1–Mirror Select	q3–Glazing Select	q3–Glazing Select
q2–Mirror	q4–Glazing A	q4–Glazing A
q3–Glazing Select	q5–Glazing B	q5–Glazing B
q4–Glazing A		
q5–Glazing B		
q6–Greenhouse		

STANDARD SIZES OF GLAZING FOR SLIDING PATIO DOORS		
Tempered 3⁄16" Single Pane	**Tempered 3⁄16" Sealed Insulating Double-Pane with 5⁄8" Thickness (1⁄4" air space)**	
46" × 76"	28" × 76"	42" × 76"
34" × 76"	33" × 74⅞"	45" × 74⅞"
	33" × 76¾"	45" × 76¾"
	34" × 76"	46" × 76"

CHAPTER 3
Walls and Floors

FRAME SPACING FOR DRYWALL CONSTRUCTION		
Single-ply Gypsum Board Thickness (in.)	**Application to Framing**	**Maximum o.c. Spacing of Framing (in.)**
Ceilings:		
3/8	Perpendicular*	16
1/2	Perpendicular	16
1/2	Parallel*	16
5/8	Parallel	16
1/2	Perpendicular*	24
5/8	Perpendicular	24
Sidewalls:		
3/8	Perpendicular or parallel	16
1/2	Perpendicular* or parallel	24
5/8	Perpendicular or parallel	24

*On ceilings to receive a water-base texture material, either hand or spray applied, install gypsum board perpendicular to framing and increase board thickness from 3/8 to 1/2 inch for 16 inch o.c. framing and from 1/2 to 5/8 inch for 24 inch o.c. framing. 3/8 inch should not support thermal insulation.

MAXIMUM FASTENER SPACING—DRYWALL

Framing	Type Const.	Type Fastener	Location	Max. Fastener Spacing			
				Gypsum Panels		Gypsum Base	
				in.	mm	in.	mm
WOOD	Single layer	Nails	Ceilings	7	178	7	178
			Sidewalls	8	203	8	203
		Screws	Ceilings	12	305	12	305
			Sidewalls	16	406	12	305
		Screws with RC-1 channels	Ceilings	12	305	12	305
			Sidewalls	12	305	12	305
	Base layer of double layer—both layers mechanically attached	Nails	Ceilings	24	610	24	610
			Sidewalls	24	610	24	610
		Screws	Ceilings	24	610	24	610
			Sidewalls	24	610	24	610
	Face layer of double layer—both layers mechanically attached	Nails	Ceilings	7	178	7	178
			Sidewalls	8	203	8	203
		Screws	Ceilings	12	305	12	305
			Sidewalls	16	406	12	305
	Base layer of double layer—face layer adhesively attached	Nails	Ceilings	7	178	7	178
			Sidewalls	8	203	8	203
		Screws	Ceilings	12	305	12	305
			Sidewalls	16	406	12	305
	Face layer of double layer—face layer adhesively attached	Nails	Ceilings	16" o.c. at ends and edges — 1 field fastener per frame member at mid-width of board	406 mm o.c. at ends and edges plus 1 field fastener per frame member	same as for gypsum panels	same as for gypsum panels
			Sidewalls	fasten top and bottom as required	fasten top and bottom as required	same as for gypsum panels	same as for gypsum panels

STEEL	Single layer	Screws	Ceilings	12	305	12	305
			Sidewalls	16	406	12	305
	Base layer of double layer—both layers mechanically attached	Screws	Ceilings	16	406	16	406
			Sidewalls	24	610	24	610
	Face layer of double layer—both layers mechanically attached	Screws	Ceilings	12	305	12	305
			Sidewalls	16	406	12	305
	Base layer of double layer—face layer adhesively attached	Screws	Ceilings	12	305	12	305
			Sidewalls	16	406	12	305
	Face layer of double layer—face layer adhesively attached	Screws	Ceilings	16" o.c. at ends and edges — 1 field fastener per frame member at mid-width of board	406 mm o.c. at ends and edges plus 1 field fastener per frame member	same as for gypsum panels	
			Sidewalls	fasten top and bottom as required	fasten top and bottom as required	same as for gypsum panels	same as for gypsum panels

DRYING TIME FOR JOINT COMPOUND UNDER TAPE

RH	RH = Relative Humidity D = Days (24 hr.) H = Hours							
98%	53 D	38 D	26 D	18 D	12 D	9 D	6 D	4½ D
97%	37 D	26 D	18 D	12 D	9 D	6 D	4½ D	3¼ D
96%	28 D	21 D	14 D	10 D	7 D	5 D	3½ D	2½ D
95%	25 D	17 D	12 D	8 D	6 D	4 D	2¾ D	2 D
94%	20 D	14 D	10 D	7 D	5 D	3¼ D	2¼ D	41 H
93%	18 D	12½ D	9 D	6 D	4 D	2¾ D	2 D	36 H
92%	15 D	11 D	8 D	5 D	3½ D	2½ D	44 H	32 H
91%	14 D	10 D	7 D	4¾ D	3¼ D	2¼ D	40 H	29 H
90%	13 D	9 D	6 D	4½ D	3 D	49 H	36 H	26 H
85%	10 D	6 D	4 D	3 D	2 D	34 H	25 H	18 H
80%	7 D	4¾ D	3¼ D	2¼ D	38 H	27 H	19 H	14 H
70%	4½ D	3½ D	2¼ D	38 H	26 H	19 H	14 H	10 H
60%	3½ D	2½ D	42 H	29 H	20 H	14 H	10 H	8 H
50%	3 D	2 D	36 H	24 H	17 H	12 H	9 H	6 H
40%	2½ D	44 H	29 H	20 H	14 H	10 H	7 H	5 H
30%	2¼ D	38 H	26 H	18 H	12 H	9 H	6 H	4½ H
20%	2 D	34 H	23 H	16 H	11 H	8 H	5½ H	4 H
10%	42 H	30 H	21 H	14 H	10 H	7 H	5 H	3½ H
0	38 H	28 H	19 H	13 H	9 H	6 H	4½ H	3 H
°F	32°	40°	50°	60°	70°	80°	90°	100°
°C	0°	4°	10°	16°	21°	27°	32°	38°

FURRING

Size of Strips	Spacing on Centers	Bd. Ft. per Sq. Ft. of Area	Nails per 1000 Bd. Ft.
1" × 2"	12" 16" 20" 24"	.18 .14 .11 .10	55 pounds
1" × 3"	12" 16" 20" 24"	.28 .21 .17 .14	37 pounds
1" × 4"	12" 16" 20" 24"	.36 .28 .22 .20	30 pounds

WALL BOARDS

Material	Size	Fastened by	Adhesive (gal.) or Nails (lb.) per 1000 Sq. Ft.
Gypsum board	48" × 96"	Nailing to studs	5 pounds
Plank T&G board	8" to 12" × 96"	Nailing to studs	2 pounds
Tempered tileboard	48" × 48"	Nailing to studs	1 pound
Tempered tileboard	48" × 48"	Adhesive to walls	1.5 gallons
Plywood panels	48" × 96"	Nailing to studs or wall	1.25 pounds
Rock lath	16" × 48"	Nailing to studs	5 pounds
Perforated hardboard	48" × 96"	Nailing to studs	4 pounds

PLASTER BOARD SPECIFICATIONS AND APPLICATIONS

Plaster Board	Thickness	Sizes	Use
Regular, paper-faced	¼"	4' × 6' to 14'	Repairing old gypsum walls.
Regular, paper-faced	⅜"	4' × 6' to 14'	Double-layer installation.
Regular, paper-faced	½" & ⅝"	4' × 6' to 14'	Standard single-ply installation.
Regular, with foil back	½" & ⅝"	4' × 6' to 14'	Use as vapor barrier, radiant heat retarder.
Type X, fire retardant	½" & ⅝"	4' × 6' to 16'	In garages, workshops, and kitchens as well as around furnaces, fireplaces, and chimney walls.
Moisture-resistant	½" & ⅝"	4' × 6' to 16'	The backing and around kitchens, spas, baths, showers, and laundry rooms.
Decorator panels	5/16"	4' × 8'	Any room.
Gypsum lath	⅜", ½", ⅝"	16" × 4'	As a base for plaster.
		2' × 8' to 12'	⅜" for 16" o.c. stud spacing, ½" or ⅝" for 24" o.c. stud spacing.

ALLOWABLE CARRYING LOADS FOR ANCHOR BOLTS

Type Fastener	Size	Allowable Load ½" Wallboard	Allowable Load ⅝" Wallboard
Hollow wall screw anchors	⅛" dia. short	50 lbs.	—
	$\frac{3}{16}$" dia. short	65 lbs.	—
	¼", $\frac{5}{16}$", ⅜" dia. short	65 lbs.	—
	$\frac{3}{16}$" dia. long	—	90 lbs.
	¼", $\frac{5}{16}$", ⅜" dia. long	—	95 lbs.
Common toggle bolts	⅛" dia.	30 lbs.	90 lbs.
	$\frac{3}{16}$" dia.	60 lbs.	120 lbs.
	¼", $\frac{5}{16}$", ⅜" dia.	80 lbs.	120 lbs.

HARDWOOD FLOORING GRADES

Flooring is bundled by averaging the lengths. A bundle may include pieces from 6 inches under to 6 inches over the nominal length of the bundle. No piece is shorter than 9 inches. Quantity with length under 4 feet held to stated percentage of total footage in any one shipment of item. ¾ inch added to face length when measuring length of each piece.

Unfinished Oak Flooring

Clear (Plain or Quarter Sawn)**
Best appearance, best grade, most uniform color, limited small character marks.
Bundles 1¼ feet and up. Average length 3¾ feet.

Select and Better (Special Order)
A combination of Clear and Select grades.

Select (Plain or Quarter Sawn)**
Excellent appearance, limited character marks, unlimited sound sap. Bundles 1¼ feet and up. Average length 3¼ feet.

No. 1 Common
Variegated appearance, light and dark colors, knots, flags, worm holes, and other character marks allowed to provide a variegated appearance, after imperfections are filled and finished. Bundles 1¼ feet and up. Average length 2¾ feet.

No. 2 Common (Red & White may be mixed)
Rustic appearance, all wood characteristics of species, a serviceable economical floor after knot holes, worm holes, checks, and other imperfections are filled and finished. Bundles 1¼ feet and up. Average length 2¼ feet.

HARDWOOD FLOORING GRADES *(cont.)*

Beech, Birch, Hard Maple

First Grade White Hard Maple (Special Order)
Same as First Grade except face all bright sapwood.

First Grade Red Beech & Birch (Special Order)
Same as First Grade except face all red heartwood.

First Grade
Best appearance, natural color variation, limited character marks, unlimited sap.
Bundles 2 feet and up; 2-foot and 3-foot bundles up to 33% footage.

Second and Better Grade
Excellent appearance, a combination of First and Second Grades.
Bundles 2 feet and up; 2-foot and 3-foot bundles up to 40% footage.
(NOTE: 5% 1¼ feet bundles allowed in Second and Better jointed flooring only.)

Second Grade
Variegated appearance, varying sound wood characteristics of species.
Bundles 2 feet and up; 2-foot and 3-foot bundles up to 45% footage.

Third and Better Grade
A combination of First, Second, and Third Grades.
Bundles 1¼ feet and up; 1¼-foot to 3-foot bundles as produced up to 50% footage.

Third Grade
Rustic appearance, all wood characteristics of species, serviceable economical floor after filling.
Bundles 1¼ foot and up; 1¼-foot to 3-foot bundles as produced up to 65% footage.

Pecan Flooring

***First Grade Red** (Special Order)
Same as First Grade except face all heartwood.

***First Grade White** (Special Order)
Same as First Grade except face all bright sapwood.

First Grade
Excellent appearance, natural color variation, limited character marks, unlimited sap.
Bundles 2 feet and up; 2-foot and 3-foot bundles up to 25% footage.

***Second Grade Red** (Special Order Only)
Same as Second Grade except face all heartwood.

Second Grade
Variegated appearance, varying sound wood characteristics of species.
Bundles 1¼ feet and up; 1¼-foot to 3-foot bundles as produced up to 40% footage.

Third Grade
Rustic appearance, all wood characteristics of species, a serviceable economical floor after filling.
Bundles 1¼ feet and up; 1¼-foot to 3-foot bundles as produced up to 60% footage.

HARDWOOD FLOORING GRADES *(cont.)*

Prefinished Oak Flooring

***Prime** (Special Order Only)
Excellent appearance, natural color variation, limited character marks, unlimited sap.
Bundles 1¼ feet and up; average length 3½ feet.

Standard and Better Grade
Combination of Standard and Prime.
Bundles 1¼ feet and up; average length 3 feet.

Standard Grade
Variegated appearance, varying sound wood characteristics of species, a sound floor.
Bundles 1¼ feet and up; average length 2¾ feet.

***Tavern and Better Grade** (Special Order Only)
Combination of Prime, Standard, and Tavern; all wood characteristics of species.
Bundles 1¼ feet and up; average length 3 feet.

Tavern Grade
Rustic appearance, all wood characteristics of species, a serviceable economical floor.
Bundles 1¼ feet and up; average length 2¼ feet.

*1¼ feet Shorts (Red & White may be mixed). Unique Variegated Appearance. Lengths 9 inches to 18 inches.
Bundles average nominal 1¼ feet. Production limited.

*No. 1 Common and Better Shorts. A combination grade, Clear, Select, and No. 1 Common, 9 inches to 18 inches.

*No. 2 Common Shorts. Same as No. 2 Common, except length 9 inches to 18 inches.

**Quarter Sawn—Special Order Only.

HARDWOOD FLOORING GRADES, SIZES, COUNTS AND WEIGHTS

"Nominal" is used by the *trade*, but it is not always the actual size, which may be 1/32 inch less than the so-called nominal size. "Actual" is the *mill* size for thickness and face width, excluding tongue width. "Counted" size determines the board feet in a shipment. Pieces less than 1 inch in thickness are considered to be 1 inch.

Nominal	Actual	Counted	Weights M Ft.
Tongue and Groove-end Matched			
**3/4 × 3 1/4"	3/4 × 3 1/4"	1 × 4"	2,210 lbs.
3/4 × 2 1/4"	3/4 × 2 1/4"	1 × 3"	2,020 lbs.
3/4 × 2"	3/4 × 2"	1 × 2 3/4"	1,920 lbs.
3/4 × 1 1/2"	3/4 × 1 1/2"	1 × 2 1/4"	1,820 lbs.
**3/8 × 2"	11/32 × 2"	1 × 2 1/2"	1,000 lbs.
**3/8 × 1 1/2"	11/32 × 1 1/2"	1 × 2"	1,000 lbs.
**1/2 × 2"	15/32 × 2"	1 × 2 1/2"	1,350 lbs.
**1/2 × 1 1/2"	15/32 × 1 1/2"	1 × 2"	1,300 lbs.
Square Edge			
**5/16 × 2"	5/16 × 2"	face count	1,200 lbs.
**5/16 × 1 1/2"	5/16 × 1 1/2"	face count	1,200 lbs.
Special Thickness (T and G, End Matched)			
**33/32 × 3 1/4"	33/32 × 3 1/4"	5/4 × 4"	2,400 lbs.
**33/32 × 2 1/4"	33/32 × 2 1/4"	5/4 × 3"	2,250 lbs.
**33/32 × 2"	33/32 × 2"	5/4 × 2 3/4"	2,250 lbs.
Jointed Flooring — i.e., Square Edge			
**3/4 × 2 1/2"	3/4 × 2 1/2"	1 × 3 1/4"	2,160 lbs.
**3/4 × 3 1/4"	3/4 × 3 1/4"	1 × 4"	2,300 lbs.
**3/4 × 3 1/2"	3/4 × 3 1/2"	1 × 4 1/4"	2,400 lbs.
**33/32 × 2 1/2"	33/32 × 2 1/2"	5/4 × 3 1/4"	2,500 lbs.
**33/32 × 3 1/2"	33/32 × 3 1/2"	5/4 × 4 1/4"	2,600 lbs.

**Special Order Only

NAIL CHART FOR APPLICATION OF STRIP FLOORING		
Flooring Nominal Size, Inches	**Size of Fasteners**	**Spacing of Fasteners**
Tongue and Groove Flooring Must Be Blind Nailed		
¾ × 1½" ¾ × 2¼" ¾ × 3¼" ¾ × 3" to 8" plank**	2" machine-driven fasteners; 7d or 8d screw or cut nail	10" – 12" apart* 8" apart into and between joists
Following Flooring Must Be Laid on a Subfloor		
½ × 1½" ½ × 2"	1½" machine-driven fastener; 5d screw, cut steel, or wire-casing nail	10" apart
⅜ × 1½" ⅜ × 2"	1¼" machine-driven fastener, or 4d bright-wire casing nail	8" apart
Square-Edge Flooring as Follows, Face-Nailed—Through Top Face		
5⁄16 × 1½" 5⁄16 × 2"	1", 15-gauge fully barbed flooring brad	2 nails every 7"
5⁄16 × 1⅓"	1", 15-gauge fully barbed flooring brad	1 nail every 5" on alternate sides of strip

*If subfloor is ½-inch plywood, fasten into each joist, with additional fastening between joists.

**Plank flooring over 4 inches wide must be installed over a subfloor.

CUTAWAY VIEW SHOWING VARIOUS LAYERS OF MATERIAL UNDER STRIP FLOORING

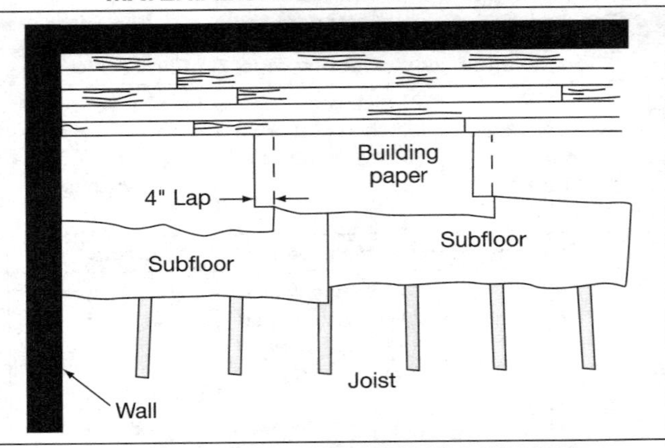

ADHESIVE FOR RESILIENT FLOOR TILE

Type and Use	Approximate Coverage in Square Feet per Gallon
Primer—For treating on or below grade concrete subfloors before installing asphalt tile.	250 to 350
Asphalt cement—For installing asphalt tile over primed concrete subfloors in direct contact with the ground.	200
Emulsion adhesive—For installing asphalt tile over lining felt.	130 to 150
Lining paste—For cementing lining felt to wood subfloor.	160
Floor and wall size—For priming chalky or dusty suspended concrete subfloors before installing resilient tile other than asphalt.	200 to 300
Waterproof cement—Recommended for installing linoleum tile, rubber, and cork tile over any type of suspended subfloor in areas where surface moisture is a problem.	130 to 150

ESTIMATING MATERIAL FOR FLOOR TILE

	Number of Tiles			
Square Footage of Floor	9" × 9"	12" × 12"	6" × 6"	9" × 18"
1	2	1	4	1
2	4	2	8	2
3	6	3	12	3
4	8	4	16	4
5	9	5	20	5
6	11	6	24	6
7	13	7	28	7
8	15	8	32	8
9	16	9	36	8
10	18	10	40	9
20	36	20	80	18
30	54	30	120	27
40	72	40	160	36
50	89	50	200	45
60	107	60	240	54
70	125	70	280	63
80	143	80	320	72
90	160	90	360	80
100	178	100	400	90
200	356	200	800	178
300	534	300	1200	267
400	712	400	1600	356
500	890	500	2000	445

Note: Add 5% for waste.

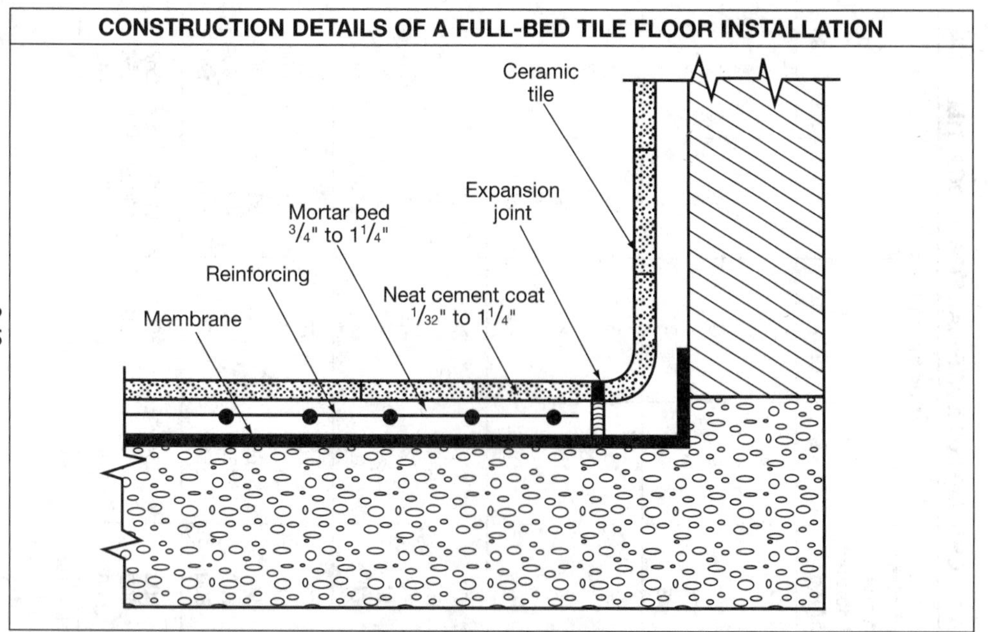
CONSTRUCTION DETAILS OF A FULL-BED TILE FLOOR INSTALLATION
Ceramic tile
Expansion joint
Mortar bed
$^3/_4$" to $1^1/_4$"
Reinforcing
Neat cement coat
$^1/_{32}$" to $1^1/_4$"
Membrane

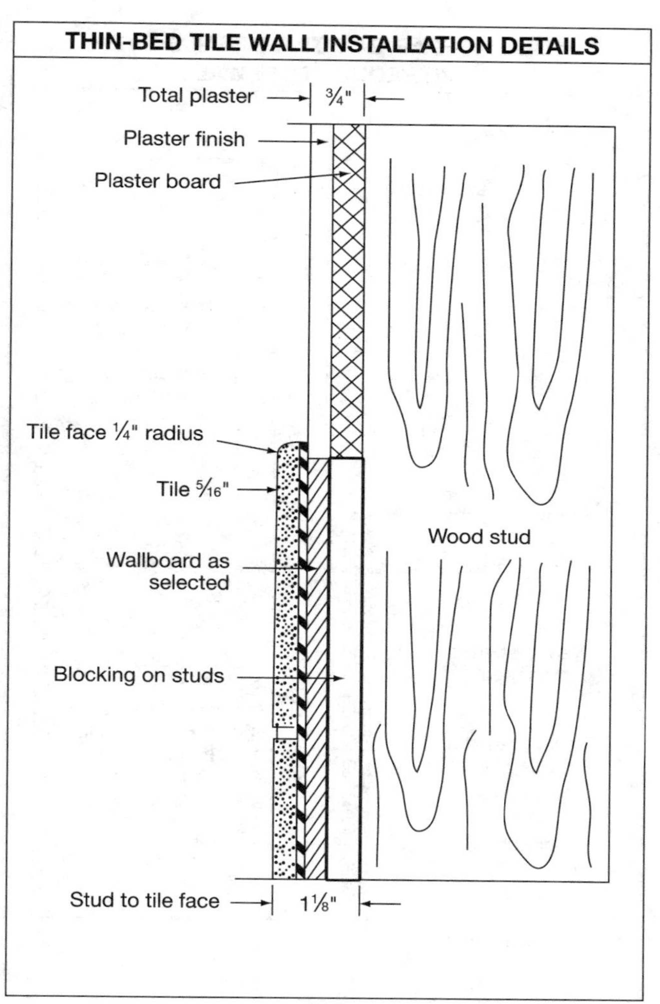
THIN-BED TILE WALL INSTALLATION DETAILS
Total plaster
¾"
Plaster finish
Plaster board
Tile face ¼" radius
Tile 5⁄16"
Wallboard as selected
Blocking on studs
Wood stud
Stud to tile face
1⅛"

FULL-MORTAR-BED TILE DETAILS OVER A VITREOUS SOLID WALL

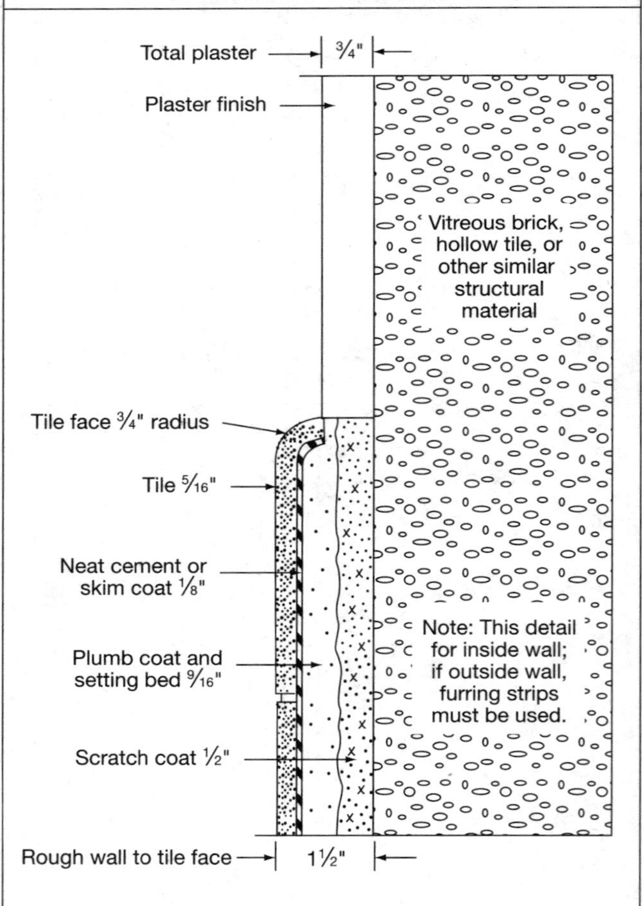

FULL-MORTAR-BED ON WOOD STUDS FOR WALL TILE

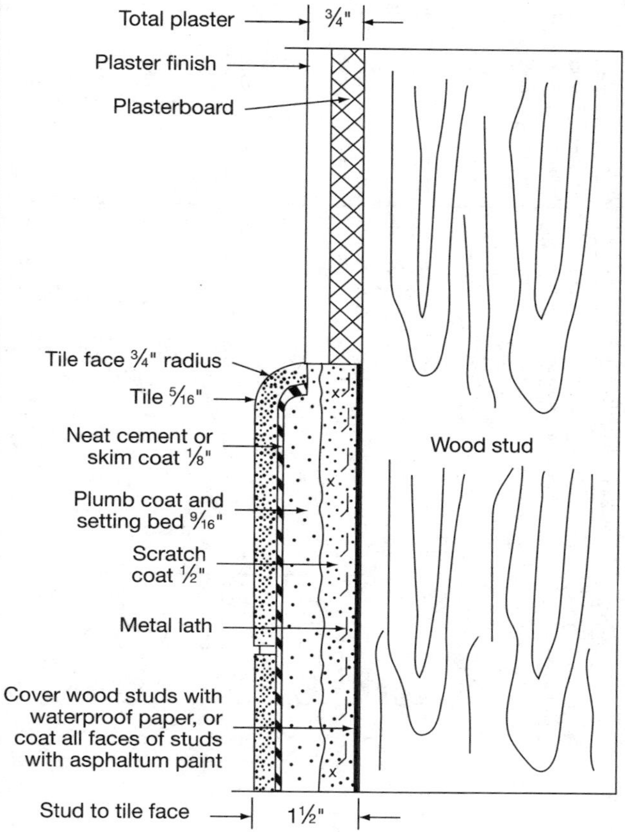

Total plaster ¾"

Plaster finish

Plasterboard

Tile face ¾" radius

Tile 5⁄16"

Neat cement or skim coat ⅛"

Plumb coat and setting bed 9⁄16"

Scratch coat ½"

Metal lath

Cover wood studs with waterproof paper, or coat all faces of studs with asphaltum paint

Wood stud

Stud to tile face 1½"

CARPET FIBER TYPES AND APPLICATIONS

Fiber	Relative Cost	Comments	Wear Life	Soil/Stain Resistance	Fade Resistance	Static Buildup*	Moisture Absorbency	Crush Resistance	Mildew
Acrylic	Medium	Wool-like finish and appearance, but fine fibered and softer, tendency to ball.	Long	Very good	Colorfast if made from colored fibers; average if dyed	Minimal	Minimal	Average	Resistant
Nylon	Wide range	Easy to clean extremely durable.	Very long	Good	Very good	High, except if not treated	Some	Good	Resistant
Polyester	Medium	Good bulk and abrasion resistance. Resists soiling, mildew, and moisture.	Long	Contact with oil-based stains. Excellent soil resistance.	Average	Minimal	Minimal	Average	Resistant
Poly-propylene Olefin	Wide range	Very durable, moisture resistant, mildew resistant, nonabsorbent, static and stain resistant.	Very long	Good	Very good	Minimal	None	Fair	Resistant

Wool	Expensive	Wool from higher altitudes is the best. Not colorfast, mildews easily, absorbs moisture, and stains.	Long	Poor	Good	Varies	High	Average	Not resistant
Silk and Wool	Very expensive	Handwoven carpets in intricate designs using wool, silk, and felt.	Long	Poor	Excellent	Varies	Some	Good	Not resistant
Cotton	Inexpensive	Used in bath mats, braided rugs, and imitation Eastern and Oriental patterns.	Average	Poor	Poor	Minimal	High	Good	Not resistant

CARPET TYPES AND APPLICATIONS

Type	Description	Applications
Level Loop	All loops the same height.	Good for high-traffic areas such as hallways. Very durable.
Multilevel Loops	Loops are made in different heights, sculptured effect.	Dens or playrooms subject to traffic, use on landings or stairs. Loops can catch on heels.
Cut Pile	Level-loop carpet in which each loop is cut into two tufts. The cut ends give a softer texture called plush. Two different effects.	Bedrooms or rarely used living rooms not subject to heavy traffic. Reasonable wear resistance.
Cut-and-Loop Pile	Multilevel sculptured effect.	Dens, living rooms, and dining rooms. Durable
Shag Carpeting	Shaggy appearance.	Play or casual areas.

SHELVING SPANS

Material	Maximum Span
¾" plywood or oriented-strand board	36"
¾" particle board or waferboard	28"
1" × 12" solid lumber	24"
½" acrylic	22"
⅜" glass	18"

CHAPTER 4
Roofing, Siding, and Insulation

TYPICAL ASPHALT ROLLS

Product	Approximate Shipping Weight		Sqs. per Pkg.	Length	Width	Side or End Lap	Top Lap	Expo-sure	UL Listing
	Per Roll	Per Square							
Mineral surface roll	75# to 90#	75# to 90#	1	36' to 38'	36"	6"	2" to 4"	32" to 34"	C
Mineral surface roll (double coverage)	55# to 70#	110# to 140#	½	36'	36"	6"	19"	17"	C
Smooth surface roll	40# to 65#	40# to 65#	1	36'	36"	6"	2"	34"	None
Saturated felt (non-perforated)	60#	15# to 30#	2 to 4	72' to 144'	36"	4" to 6"	2" to 19"	17" to 34"	None

TYPICAL ASPHALT SHINGLES

Product	Configuration	Per Square			Size		Exposure	UL Listing
		Approx. Shipping Weight	Shingles	Bundles	Width	Length		
Self-sealing random-tab strip shingle Multi-thickness	Various edge, surface-texture, and application treatments	285# to 390#	66 to 90	4 or 5	$11^1/_2$" to 14"	36" to 40"	4" to 6"	A or C Many wind resistant
Self-sealing random-tab strip shingle Single-thickness	Various edge, surface-texture, and application treatments	250# to 300#	66 to 80	3 or 4	12" to $13^1/_4$"	36" to 40"	5" to $5^5/_8$"	A or C Many wind resistant

Self-sealing square-tab strip shingle Three-tab	Two-tab or Four-tab	215# to 325#	66 to 80	3 or 4	12" to 13$^1/_4$"	36" to 40"	5" to 5$^5/_8$"	A or C All wind resistant
	Three-tab	215# to 300#	66 to 80	3 or 4	12" to 13$^1/_4$"	36" to 40"	5" to 5$^5/_8$"	
Self-sealing square-tab strip shingle No-cutout	Various edge and surface-texture treatments	215# to 290#	66 to 81	3 or 4	12" to 13$^1/_4$"	36" to 40"	5" to 5$^5/_8$"	A or C All wind resistant
Individual interlocking shingle Basic design	Several design variations	180# to 250#	72 to 120	3 or 4	18" to 22$^1/_4$"	20" to 22$^1/_2$"	—	C Many wind resistant

WOOD SHINGLES

Laid to Weather	Shingles per 100 Sq. Ft.	Waste	Shingles per 100 Sq. Ft. with Waste	Nails per 100 Sq. Ft.	
				3d Nails	4d Nails
4"	900	10%	990	$3\frac{3}{4}$ pounds	$6\frac{1}{2}$ pounds
5"	700	10%	792	3 pounds	$5\frac{1}{4}$ pounds
6"	600	10%	660	$2\frac{1}{2}$ pounds	$4\frac{1}{4}$ pounds

NAIL REQUIREMENTS FOR ASPHALT ROOFING

Type of Roofing	Shingles per Square	Nails per Shingle	Length of Nail	Nails per Square	Pounds per Square	
					12-gauge by $\frac{7}{16}$" head	11-gauge by $\frac{7}{16}$" head
Roll roofing on new deck	–	–	1"	252	.73	1.12
Roll roofing over old roofing	–	–	$1\frac{3}{4}$"	252	1.13	1.78
19" selvage over old shingle	–	–	$1\frac{3}{4}$"	181	.83	1.07
3-tab sq. butt on new deck	80	4	$1\frac{1}{4}$"	336	1.22	1.44
3-tab sq. butt reroofing	80	4	$1\frac{3}{4}$"	504	2.38	3.01
Hex strip on new deck	86	4	$1\frac{1}{4}$"	361	1.28	1.68
Hex strip reroofing	86	4	$1\frac{3}{4}$"	361	1.65	2.03
Giant American	226	2	$1\frac{1}{4}$"	479	1.79	2.27
Giant Dutch lap	113	2	$1\frac{1}{4}$"	236	1.07	1.39
Individual hex	82	2	$1\frac{3}{4}$"	172	.79	1.03

ROOFING MATERIALS

Material	Life	Fire Rating	Roof Pitch	Notes:
Asphalt-Fiberglass Composite Shingle	15 to 40 years	A	4/12 or greater	Can be walked on. Economical, durable, easy application, and low maintenance. Apply at temperatures above 50°F. Can blow off in high winds. Susceptible to mildew and moss.
Asphalt Organic Felt-base Shingles	12 to 20 years	C	4/12 or greater	Can be walked on. Economical, durable, easy application, and low maintenance. Less durable and fire-resistant than asphalt-fiberglass composite. Can blow off in high winds. Susceptible to mildew and moss.
Asphalt Roll Roofing	5 to 7 years	Varies from A to C depending on coating.	1/12 or greater	Economical Easy application May have poor fire rating.

ROOFING MATERIALS *(cont.)*

Material	Life	Fire Rating	Roof Pitch	Notes:
Wood Shingles and Shakes	20 to 40 years	Varies from A to C depending on coating	Shakes: 4/12 or greater Shingles: 3/12 or greater	Natural appearance, durable. Flammable if not treated. May attract insects. High maintenance. Subject to mold and mildew. Difficult application.
Concrete Tile	over 50 years	A	3/12 or greater	Extremely durable, fireproof. High cost. Can crack. Cannot be used in freeze/thaw areas. Heavier framing may be needed.
Clay Tile	over 50 years	A	3/12 or greater	(Same as above.)
Concrete Tile	30 to 50 years	A or B	3/12 or greater	Extremely durable, fireproof or fire-resistant. Moderate cost. Cannot be used in freeze/thaw areas.
Slate Tile	over 50 years	A	4/12 or greater	Extremely durable, fireproof. Very high cost. Heavier framing may be needed. High maintenance, cannot be walked on.

Metal Shingles (Aluminum)	over 50 years	Varies from A to C	4/12 or greater	Low maintenance, can be applied over existing roofing. Fire resistant. Prone to scratches or dents. May be difficult to install and expensive. Must be insulated from noncompatible materials.
Metal Shingles (Steel)	over 50 years	Varies from A to C	3/12 or greater	Low maintenance, can be applied over existing roofing. Fire resistant. Prone to scratches or dents. Complicated to install. Can be expensive.
Metal Panels (Steel)	20 to 50 years	Varies from A to C	1/12 or greater	Easy application, durable, low maintenance. Complicated to install.

ROOF SYSTEMS AND INSTALLATION

Roofing	Attachment	Typical Underlayment	Normal Slope (in., rise/run)	Minimum Slope (in., rise/run)
Asbestos Cement Shingles	Asphalt shingle nails	Single layer, No. 15 or No. 30 asphalt-saturated felt	5/12	3/12
Asphalt Shingles	Asphalt shingle nails, staples	Single layer, No. 15 asphalt-saturated felt	4/12	2/12
Fiberglass Shingles	Asphalt shingle nails, staples	Single layer, No. 15 asphalt-saturated felt	4/12	2/12
Wood Shakes	2 nails per shake only	No. 30 asphalt-saturated felt underlayment starter course at eaves and interlayment between each shake course over entire roof	4/12	3/12
Wood Shingles	3d for 16" and 18" shingles; 4d for 24" shingles 5d or 6d for reroofing	None	5/12	3/12
Roll Roofing and Asphalt	Plastic roofing cement for flashing lap, asphalt cement for overlaps	None	3/12	Flat

Granulated Roofing Shear	Plastic roofing cement for flashing; lap cements for overlap	None	3/12	2/12
Resiply	Under-surface remelt bonding	None	Flat or 3/12	Flat
Slate	Zinc-coated or copper 3d nails for 18" slate; 4d for 20" or longer, 6d for hips and ridges. Cover nail heads with elastic cement compound.	No. 30 asphalt-saturated felt	5/12	4/12
Tile	Mission tile installed over 1½" furring strips nailed to sheathing vertically. For English tile use horizontal furring strips.		5/12	4/12

GUIDE TO ROOFING MATERIAL APPLICATIONS

Sheathing	Notes:
Slat Sheathing	1 × 4 or 1 × 6 lumber. Used with wood shingle, wood shake, concrete tile, clay tile, or metal panels.
Plywood Sheathing	Performance-rated panels. Usually ½" thickness Exterior, Exposure 1, or Exposure 2 may be used. Use with asphalt shingles, asphalt composite shingles, asphalt roll roofing, slate shingles, or metal shingles. Stagger panels on 4-foot centers. Leave ⅛" gap for expansion between panels. Use panel clips on edges between panels.
OSB Sheathing	Performance-rated panels. Usually ½" thickness Exterior, Exposure 1, or Exposure 2 may be used. Use with asphalt shingles, asphalt composite shingles, asphalt roll roofing, slate shingles, or metal shingles. Stagger panels on 4-foot centers. Leave ⅛" gap for expansion between panels. Use panel clips on edges between panels.
Tongue-and-Groove Sheathing	2 × 6 tongue-and-groove roof decking. Use where the sheathing will be visible inside the house. Stagger joints and leave 1/16" for expansion. Can be used with all roofing materials.
Underlayment	**Notes:**
Roofing Felt	Used with asphalt shingles, asphalt composite shingles, asphalt roll roofing, wood shakes, wood shingles, concrete tile, clay tile, and slate. Sold in rolls that cover 2, 4, or 5 squares. Normal weights are 15, 30, and 45 pounds per square. 30 pounds per square is most common. Two or more overlapping layers are normally required.

PLYWOOD ROOF SHEATHING

Recommended Thickness	Maximum Spacing of Supports (C. to C.)			Nail Size	Nail Spacing	
	20PSF	30PSF	40PSF		Panel Edge	Intermediate
5/16"	20"	20"	20"	6d common	6"	12"
3/8"	24"	24"	24"	6d common	6"	12"
1/2" (a)	32"	32"	30"	6d common	6"	12"
5/8" (a)	42"	42"	39"	8d common	6"	12"
3/4" (a)	48"	47"	42"	8d common	6"	12"

[a]Provide blocking or other means of suitable edge support when span exceeds 28" for 1/2" or 32" for 3/4".

INSULATING SHEATHING

Walls

Galvanized roofing nails with ¾" diameter heads (or 16-gauge wire staples with ¾" crown).

Length: sheathing thickness plus ¾"

(Length of staple legs: sheathing thickness plus ½")

Fasteners on 24" center on both field and perimeter

(4' × 8', 24" O.C.—15 nails)

(4' × 8', 16" O.C.—20 nails)

(4' × 8', 24" O.C.—15 nails)

Frame Roof

Galvanized roofing nails with ⅜" diameter heads.

Length: sheathing thickness plus ¾"

Fasteners on 24" centers on both field and perimeter

(4' × 8', 24" O.C.—15 nails)

(4' × 8', 16" O.C.—20 nails)

Interior Partitions

Secure sheathing to studs with galvanized roofing nails with ⅜" heads.

Length: sheathing thickness plus ½" (minimum)

Fasteners on 24" center on both field and perimeter

(4' × 8', 24" O.C.—15 nails)

Partitions: fasten 8" O.C. Add thickness of gypsum to thickness of sheathing plus ¾" (minimum).

Ceilings: Fasten 7" O.C. Add thickness of gypsum to thickness of sheathing plus ¾" (minimum).

SHINGLE NAILS

Application	Threading	Nail Sizes
Strip or individual shingle (new construction)	Plain or helix	1¼"
Over asphalt roofing (reroofing)	Helix or annular	1½"
Over wood shingles (reroofing)	Helix or barbed	1¾"

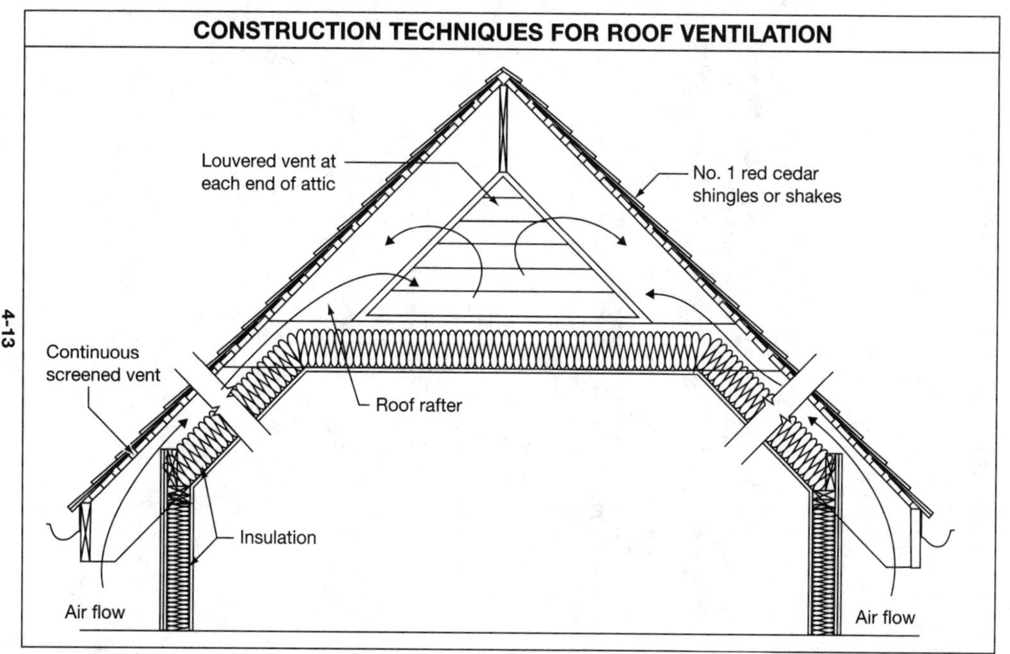
CONSTRUCTION TECHNIQUES FOR ROOF VENTILATION
Louvered vent at
each end of attic
No. 1 red cedar
shingles or shakes
Continuous
screened vent
Roof rafter
Insulation
Air flow
Air flow

GUTTER AND DOWNSPOUT SIZING

Roof Area	Gutter Size	Downspout Size
Up to 750 sq. ft.	4"	3"
750 to 1,400 sq. ft.	5"	4"
Over 1,400 sq. ft.	6"	4"

EXPOSED PLYWOOD PANEL SIDING

Minimum Thickness[a]	Minimum No. of Plies	Stud Spacing (inches) Plywood Siding Applied Direct to Studs or Over Sheathing
3/8"	3	16[1]
1/2"	4	24

Thickness of grooved panels is measured at bottom of grooves.

[a]May be 24 inches if plywood siding applied with face grain perpendicular to studs or over one of the following:

1. 1-inch board sheathing.
2. 15/32-inch plywood sheathing or
3. 3/8-inch plywood sheathing with face grain of sheathing perpendicular to studs.

ALLOWABLE SPANS FOR EXPOSED PARTICLEBOARD PANEL SIDING

Grade	Stud Spacing (inches)	Minimum Thickness (inches)		
		Siding		Exterior Ceilings and Soffits
		Direct to Studs	Continuous Support	Direct to Supports
2-M-W	16	3/8	5/16	5/16
	24	1/2	5/16	3/8
2-M-1	16	5/8	3/8	—
2-M-2				
2-M-3	24	3/4	3/8	—

BEVEL SIDING

Siding for 100 Sq. Ft. Wall

Size	Exposed to Weather	Add for Lap	Bd. Ft. per 100 Sq. Ft.	Nails per 100 Sq. Ft.
½" × 4"	2¾"	46%	151	1½ pounds
½" × 5"	3¾"	33%	138	1½ pounds
½" × 6"	4¾"	26%	131	1 pound
½" × 8"	6¾"	18%	123	¾ pound
⅝" × 8"	6¾"	18%	123	¾ pound
¾" × 8"	6¾"	18%	123	¾ pound
⅝" × 10"	8¾"	14%	119	½ pound
¾" × 10"	8¾"	14%	119	½ pound
¾" × 12"	10¾"	12%	117	½ pound

NOTE: Quantities include 5 percent for end-cutting and waste. Deduct for all openings over 10 square feet.

DROP SIDING

Siding for 100 Sq. Ft. Wall

Size	Exposed to Weather	Add for Lap	Bd. Ft. per 100 Sq. Ft.	Nails per 100 Sq. Ft.
1" × 6"	5¼"	14%	119	2½ pounds
1" × 8"	7¼"	10%	115	2 pounds

NOTE: Quantities include 5 percent for end-cutting and waste. Deduct for all openings over 10 square feet.

FASTENING FOR WALL SHEATHING, PANEL SIDING, AND FLOOR UNDERLAYMENT

Spacing Specifications (in inches)				Fastener Specifications	
Material	**Thickness**	**Edges**	**Intermediate**	**Leg Length (in inches)**	**Fastener**
Plywood	3/8"	6	12	1 1/2	6d galv. casing nail *or* 6d galv. siding nail
Panel	1/2"	6	12	1 5/8	6d galv. casing nail *or* 6d galv. siding nail
Siding 2 + 3	5/8"	6	12	1 7/8	8d galv. casing nail *or* 8d galv. siding nail
Fiberboard Wall Sheathing	1/2"	6 4	12 10	1 1/2	No. 14 gauge staple *or* No. 15 gauge staple *or* No. 16 gauge staple
	25/32"	6 4	12 10	1 3/4	No. 14 gauge staple *or* No. 15 gauge staple *or* No. 16 gauge staple
Gypsum Wall Sheathing	1/2"	6 4	12 10	1 1/2	No. 14 gauge staple *or* No. 15 gauge staple *or* No. 16 gauge staple
Floor Underlayment: plywood, hardboard, flakeboard, particle board	1/4" & 5/16"	6 4	8-grid 6-grid	1 1/4 7/8	3d ring-shank nail *or* No. 19 gauge staple (3/16" crown)
	3/8"	6 4	8-grid 6-grid		3d ring-shank nail *or* No. 16 gauge staple
	1/2"	6 4	8-grid 8-grid	1 1/4	6d box nail *or* 3d ring-shank nail *or* No. 16 gauge staple
	5/8"	6 4	8-grid 6-grid		6d box nail *or* 4d ring-shank nail *or* No. 16 gauge staple

STAPLING WOOD MEMBERS TO PLYWOOD WALL SHEATHING				
Nominal Thickness (inches)	Spacing Specifications (in inches)		Fastener Specifications	
	Edges	Intermediate	Leg Length (in inches)	Fastener
1/4 & 5/16	6	12	1 3/8	14-gauge staple *or* 15-gauge staple *or* 16-gauge staple
			1 1/2	6d box nail
3/8	6	12	1 3/8	14-gauge staple *or* 15-gauge staple *or* 16-gauge staple
			1 5/8	6d common nail *or* 6d ring-shank nail *or* 6d screw-shank nail
			1 1/2	6d box nail

STAPLING FOR PLYWOOD SUBFLOOR AND ROOF SHEATHING

Subfloor or Sheathing Nominal Thickness (in inches)	Spacing Specifications (in inches)			Leg Length (in inches)	Fastener
	Intermediate Edges	Roof	Floor		
$\frac{7}{8}$	6	12	10	$2\frac{3}{8}$	8d common *or* 8d ring-shank nail *or* 8d screw-shank nail
				2 $2\frac{1}{8}$	13-gauge staple 14-gauge staple
	4	10	8	$2\frac{1}{8}$ $2\frac{1}{2}$	15-gauge staple 16-gauge staple
1	6	12	10	$2\frac{1}{8}$	13-gauge staple
				$2\frac{1}{3}$	8d common nail *or* 8d ring-shank nail *or* 8d screw-shank nail
				$2\frac{1}{4}$	14-gauge staple
	4	10	8	$2\frac{3}{8}$	15-gauge staple
$1\frac{1}{8}$–$1\frac{3}{4}$	6	12	10	$2\frac{5}{8}$	13-gauge staple
				$2\frac{7}{8}$	10d common nail *or* 14-gauge staple
				$2\frac{1}{2}$	8d ring-shank nail *or* 8d screw-shank nail
	4	10	8	3	15-gauge staple
$\frac{1}{4}$–$\frac{5}{16}$–$\frac{3}{8}$	6	12	10	$1\frac{3}{8}$	14-gauge staple *or* 15–gauge staple
				$1\frac{5}{8}$	16-gauge staple *or* 6d common nail *or* 6d ring-shank nail *or* 6d screw-shank nail

STAPLING FOR PLYWOOD SUBFLOOR AND ROOF SHEATHING *(cont.)*

Subfloor or Sheathing Nominal Thickness (in inches)	Spacing Specifications (in inches)			Leg Length (in inches)	Fastener
	Intermediate Edges	Roof	Floor		
1/2	6	12	10	1 1/2	14-gauge staple *or* 15-gauge staple
				1 3/4	16-gauge staple *or* 6d common nail *or* 6d ring-shank nail *or* 6d screw-shank nail
5/8	6	12	10	1 7/8	14-gauge staple
				2	15-gauge staple
				2 1/4	16-gauge staple
				2 1/8	8d common nail
				1 7/8	6d ring-shank nail *or* 6d screw-shank nail
3/4	6	12	10	2	14-gauge staple
				2 1/8	15-gauge staple
	6	10	8	2 3/8	16-gauge staple
	6	12	10	2 1/4	8d common nail
				2	6d ring-shank nail *or* 6d screw-shank nail

ONE-HOUR EXTERIOR WALL CONSTRUCTION

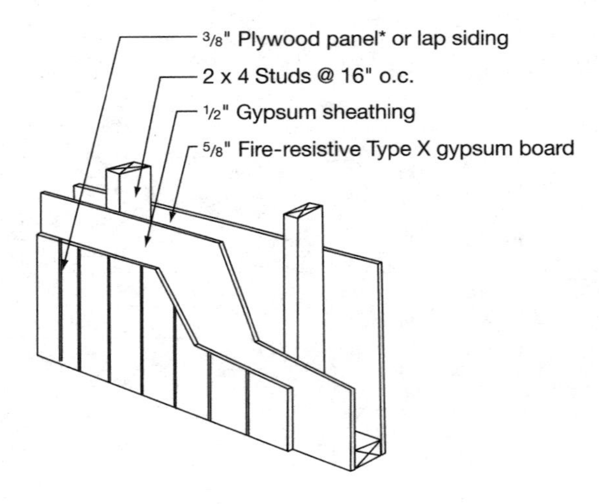

*Including nominal 3/8" specialty plywood sidings

JOINT DETAILS FOR PLYWOOD SIDING

Vertical Wall Joints

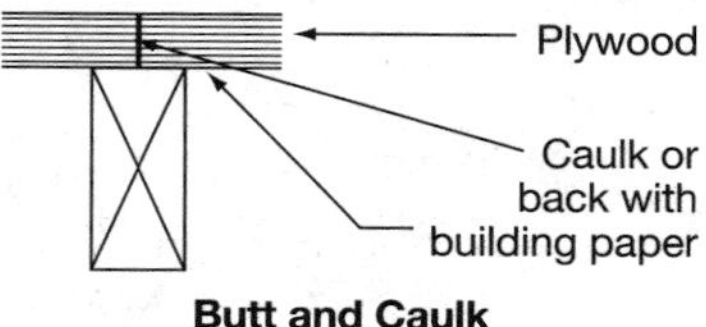

Butt and Caulk

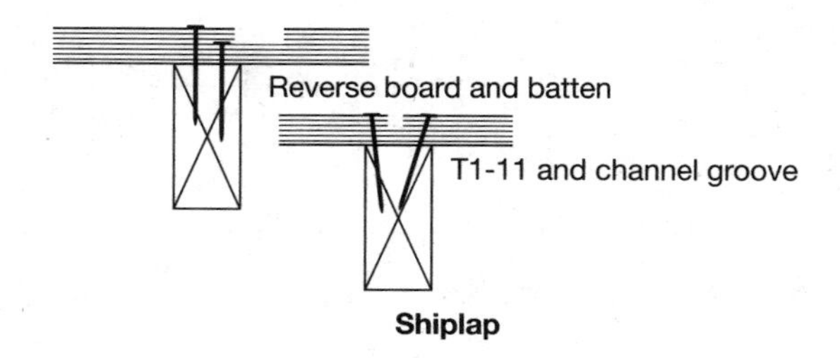

Shiplap

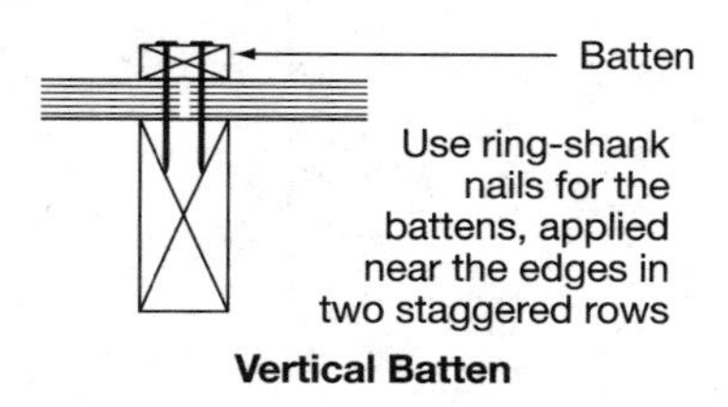

Vertical Batten

JOINT DETAILS FOR PLYWOOD SIDING *(cont.)*

Vertical Inside and Outside Corner Joints

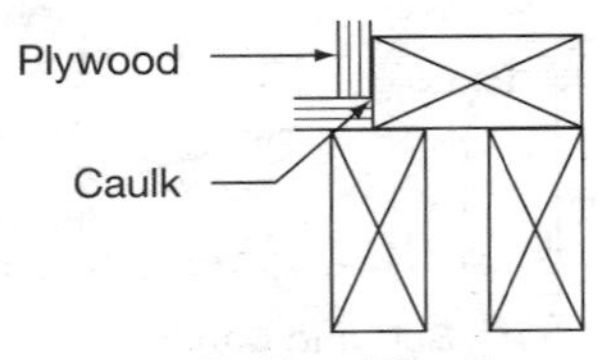

Butt and Caulk

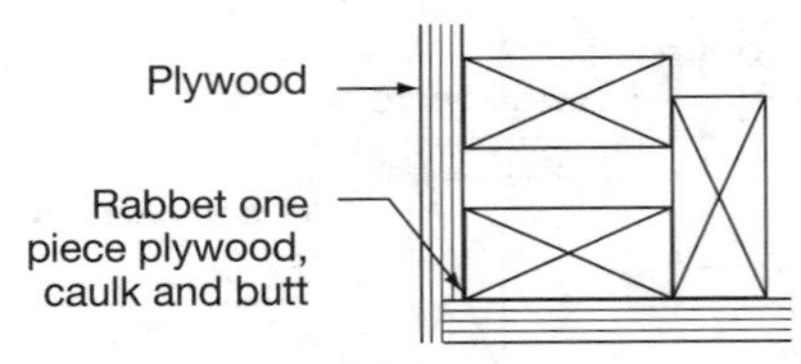

Rabbet and Caulk

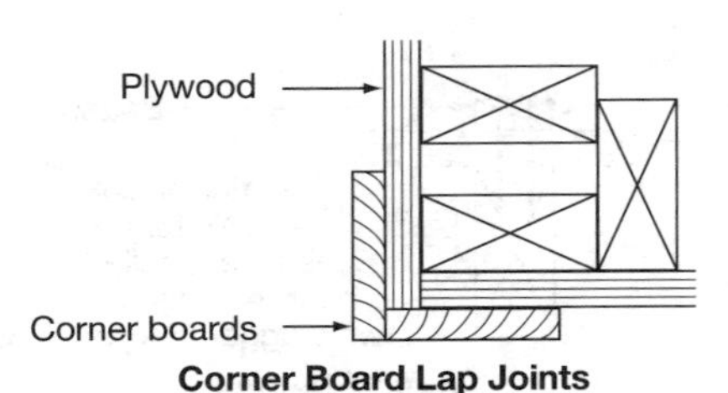

Corner Board Lap Joints

JOINT DETAILS FOR PLYWOOD SIDING *(cont.)*

Horizontal Wall Joints

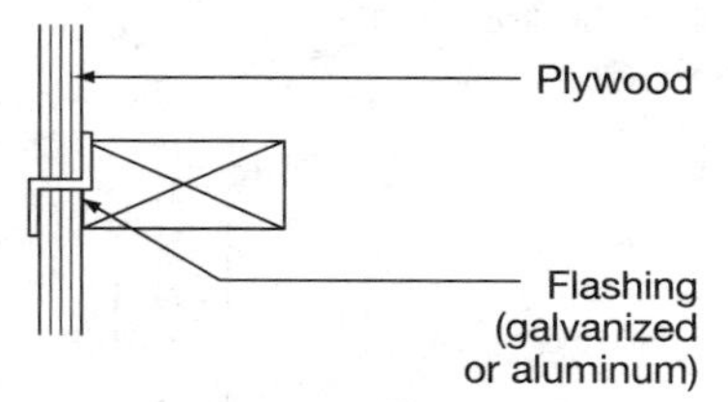

Butt and Flash

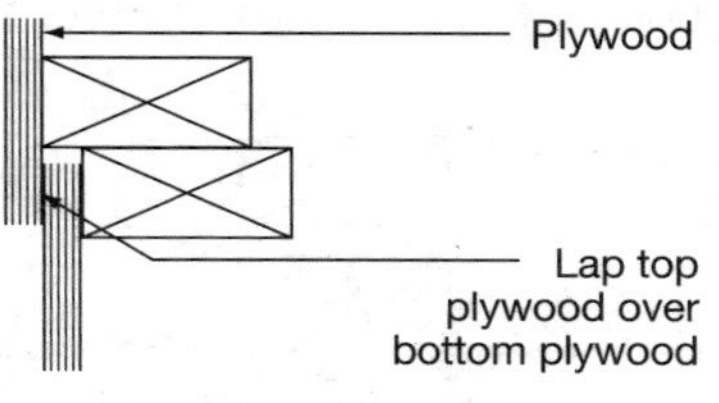

Lap Plywood

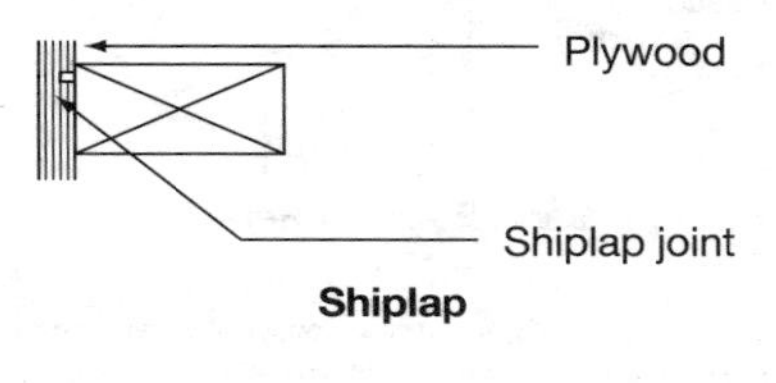

Shiplap

JOINT DETAILS FOR PLYWOOD SIDING *(cont.)*

Horizontal Beltline Joints

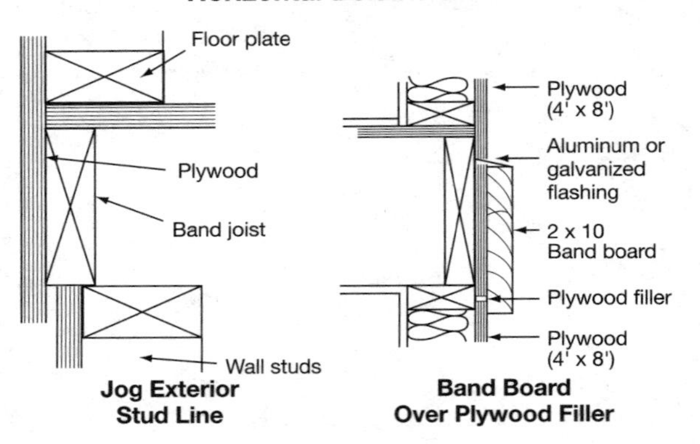

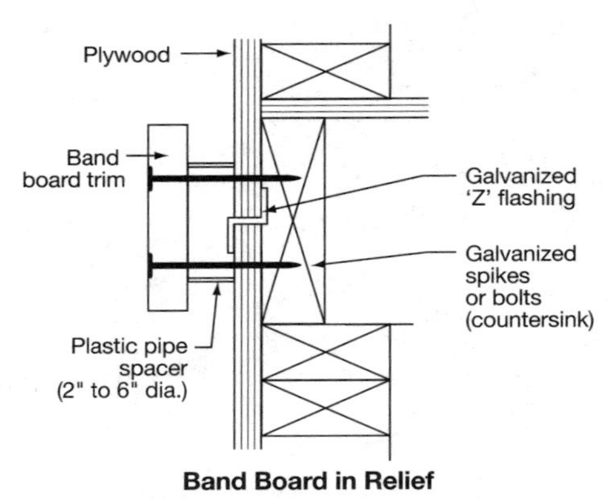

Band Board in Relief

Note: For multi-story buildings, make provisions at horizontal joints for settling shrinkage of framing, especially when applying siding direct to studs.

JOINT DETAILS FOR PLYWOOD SIDING *(cont.)*

Window Details

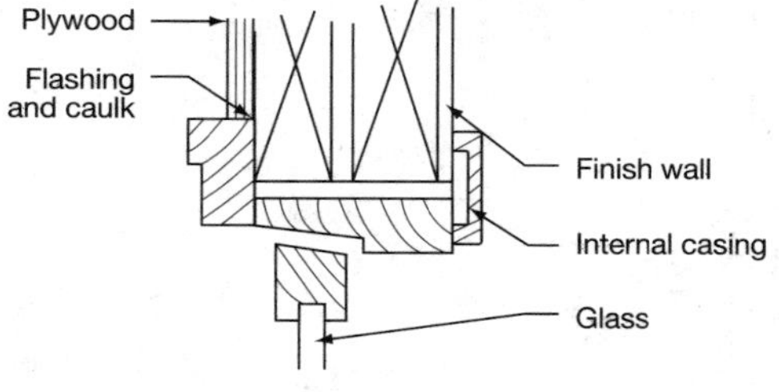

Head

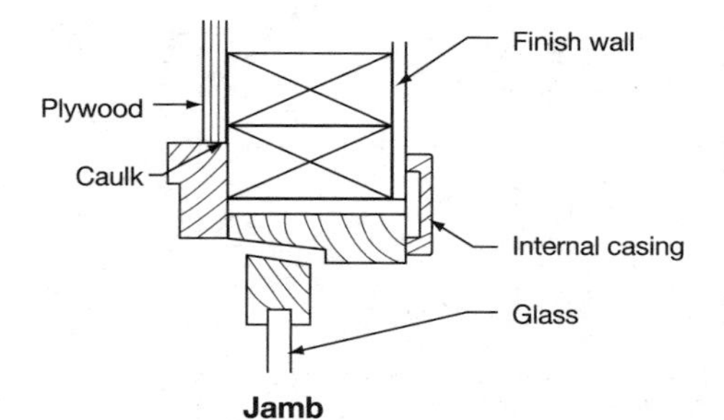

Jamb

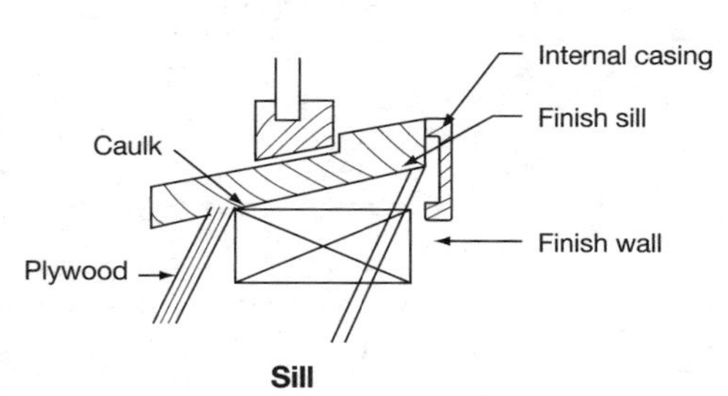

Sill

STUCCO FORMULAS

Coat	Thickness	Mix	Parts by Volume					
			Portland Cement	Lime	Masonry Cement	Volume per Sum of Cementitious Materials: Sand	or Perlite	Water
1st (Scratch Coat)	3/8"	A	1		1–2	2½–4	1½–2	Add and mix clean water to desired consistency.
		B	1	0–¾		2¼–4	3–4	
2nd (Brown Coat)	3/8"	A	1		1–2	3–5	1½–2	
		B	1	0–¾		3–5	3–4	
3rd (Finish Coat)	3/8"	A	1		1	3	2	
		B	1	¾–1½		3	2	

Notes: Start with mix A for first coat, then use the remaining A mix for the 2nd and 3rd coats.
Maximum batch life is 2 to 2½ hours.
Don't use ribbed lath on exterior.
All lath must be rust and oil free.
Use coarse sand.
Keep each coat moist three to five days before applying next coat.
After top coat, allow at least two weeks before painting. Use alkali-resistant paint.

STUCCO LATH USES BY CATEGORY	
Type	**Application**
Expanded Metal	Zinc coated steel with backing paper and furring fasteners.
Stucco Mesh	Use on solid surfaces, needs backing paper and furring fasteners.
Diamond Mesh (self-furring)	Use 3.4 weight or heavier.
Flat Rib Lath	In $\frac{1}{8}$", $\frac{3}{8}$" and $\frac{3}{4}$" only. Use 3.4 weight or heavier.
Wire Lath	Galvanized steel wire.
Welded Wire	18 gauge for 1" × 1" squares or 16 gauge for 2" × 2" openings.
Welded Wire (backed)	18 gauge for 1" × 1" squares or 17 gauge for 1" × $\frac{1}{2}$" openings.
Woven Wire Fabric	16 gauge for 2" × 2" openings and 18 gauge for 1" × 1" openings.

MINIMUM R-VALUES FOR U.S. CLIMATE ZONES

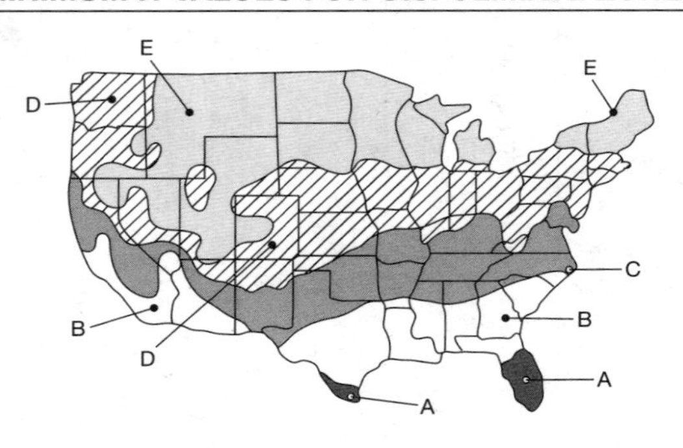

OIL HEAT, GAS HEAT, OR HEAT PUMPS

	Zone A	Zone B	Zone C	Zone D	Zone E
Ceiling Insulation	R-19	R-19	R-26	R-30	R-38
Wall Insulation	R-11	R-11	R-13	R-13	R-19
Floors over			R-11	R-11	R-19
Unheated Spaces	R-11	R-13	R-19	R-19	R-19
Foundation Walls of Heated Spaces	none	none	R-6	R-11	R-11
Slab Foundation Perimeter	none	R-2	R-5	R-5	R-7.5

ELECTRIC RESISTANCE HEAT

	Zone A	Zone B	Zone C	Zone D	Zone E
Ceiling Insulation	R-19	R-22	R-30	R-30	R-38
Wall Insulation	R-11	R-13	R-19	R-19	R-19
Floors over Unheated Spaces	none	R-11	R-19	R-19	R-19
Foundation Walls of Heated Spaces	none	none	R-6	R-11	R-11
Slab Foundation Perimeter	none	R-5	R-7.5	R-7.5	R-7.5

INSULATION PROPERTIES

Insulation	Form	Product Thickness	R-value	Wood Frame Wall: Attic	Wood Frame Wall: Interior Face	Wood Frame Wall: Exterior Face	Wood Frame Wall: Stud Space	Masonry Walls Aboveground: Exterior	Masonry Walls Aboveground: Interior	Masonry Walls Aboveground: Interior or Crawl Space	Crawl Space/Basement: Under Slab	Crawl Space/Basement: Basement Floor Between Sleepers	Crawl Space/Basement: Floors Over Unheated Space
Fiberglass and Mineral Wool	Attic blanket	8"	R-25	•			•		•	•		•	•
	Batts,	12"	R-38	•			•		•	•		•	•
	flat	9½"	R-30	•			•		•	•		•	•
	Rolled	6¾"	R-22	•			•		•	•		•	•
	batts	6¼"	R-19	•			•		•	•		•	•
		3⅝"	R-13	•			•		•	•		•	•
		3½"	R-11	•			•		•	•		•	•
		6¼"	R-19	•			•		•	•		•	•
		3½"	R-11	•			•		•	•		•	•
Polyurethane, Polyisocyanurate	Foam board	½"	R-3.6	•	•	•		•	•	•	•	•	
		⅝"	R-4.5	•	•	•		•	•	•	•	•	

INSULATION PROPERTIES *(cont.)*

Insulation	Form	Product Thickness	R-value	Wood Frame Wall: Attic	Wood Frame Wall: Interior Face	Wood Frame Wall: Exterior Face	Wood Frame Wall: Stud Space	Masonry Walls Aboveground: Exterior	Masonry Walls Aboveground: Interior	Masonry Walls Aboveground: Interior or Crawl Space	Crawl Space/ Basement: Under Slab	Crawl Space/ Basement: Basement Floor Between Sleepers	Crawl Space/ Basement: Floors Over Unheated Space
Polystyrene, molded or expanded	Foam board	3/4"	R-5.4	•	•	•		•	•	•	•	•	
		7/8"	R-6.3	•	•	•		•	•	•	•	•	
		1"	R-7.2	•	•	•		•	•	•	•	•	
		1 1/4"	R-9	•	•	•		•	•	•	•	•	
		1 1/2"	R-10.8	•	•	•		•	•	•	•	•	
		1 3/4"	R-12.6	•	•	•		•	•	•	•	•	
		2"	R-14.4	•	•	•		•	•	•	•	•	
		3/4"	R-3		•	•		•		•	•		
		1"	R-4		•	•		•		•	•		
		1 1/2"	R-6		•	•		•		•	•		
		2"	R-8		•	•		•		•	•		
		3"	R-12		•	•		•		•			

INTERIOR BOARD INSULATION PROPERTIES

	Polyisocyanurate Polyurethane (Foil Faced)	Polystyrene Extruded	Glass Fiberboard	Polystyrene Molded or Expanded	Phenolic Foamboard
Use					
Frame: exterior wall sheathing	•	•			
Exterior basement/crawl space insul.		•	•		
Exterior masonry wall above ground	•	•	•	•	•
Interior masonry wall	•	•	•	•	•
Interior frame wall	•	•	•	•	
Cathedral ceiling	•	•	•	•	
Attic floor					
Basement floor	•	•	•		
Widths and Lengths	4' × 8', 4' × 9'	4' × 8', 4' × 9'	2' × 4', 4' × 8', 4' × 9'	2' × 8', 4' × 8', 4' × 9'	
Facings	Foil-kraft foil both sides; Foil-kraft foil one side, foil other	Foil, foil-kraft foil, and film facings	Unfaced	Film facings, Foil, Foil-kraft foil, film or unfaced	
Shrinkage	Some	None	None	Some	
Moisture absorption		None	None	Some	
Corrosion	Foil protects	None	None	None	
Fungal-bacterial growth	Does not support growth	Does not support growth	Does not promote growth	Does not support growth	
UV degradation	Foil protects	Yes	None	Yes	
Other	Easy to cut and install	Easy to cut and install	Easy to cut and install	Easy to cut and install	Tends to crumble—use screw installation rather than nailing

R VALUES FOR FRAME CONSTRUCTION							
Wall R-Value	Application	Description	Insulation Type	Insulation Layers	Thickness Each (inches)	R Value Each	R Value Achieved
R-40	Double wall	2 walls of 2" × 4" studs on 16" or 24" centers, spaced 3½" apart	Fiberglass or rock wool batts, blankets, or rolls	3 layers (2 vertical, 1 horizontal)	3⅝	13	39
			Fiberglass or rock wool batts, blankets, or rolls and polyisocyanurate foamboard sheathing	2 layers Exterior Interior 1 layer sheathing	 3⅝ 6¼ ¾	 13 19 5.4	 36.4
R-30	Double wall	2 walls of 2" × 4" studs on 16" or 24" centers spaced 2½" apart	Fiberglass or rock wool batts, blankets, or rolls	1 layer 1 layer	6¼ 3½	19 11	30

	Single wall	2" × 6" × 24" o.c. wall with horizontal 2" × 4" strapping 24" o.c.	Fiberglass or rock wool	1 layer 1 layer (between strapping)	6¼ 3½	19 11	29
		2" × 6" × 24" o.c. wall	Fiberglass or rock wool and polyisocyanurate foam-board sheathing	1 layer 1 layer sheathing	6¼ 1½	19 10.8	28.8
R-26	Single wall	2" × 6" × 24" o.c. wall	Fiberglass or rock wool and polyisocyanurate foam-board sheathing	1 layer 1 layer sheathing	6¼ ¾	19 5.4	23.4

R VALUES FOR GYPSUM CEILINGS

Insulation Form	Allowable R Values		
	½" Gypsum 24" o.c.	½" Gypsum 16" o.c.	⅝" Gypsum 24" o.c.
Blown or Poured:			
Cellulose	R-22	R-37	R-37
Rock wool	R-25	R-43	R-43
Glass fiber	R-51	R-87	R-87
Glass fiber	R-56	R-95	R-95
Batts, Rolls, or Blankets:			
Glass fiber batts	R-95+	R-95+	R-95+
Rock wool batts	R-25	R-42	R-42

MINIMUM RESIDENTIAL PIPE INSULATION THICKNESSES (in inches)

Pipe System	Fluid Temperature Range, °F	1" & less	Pipe Sizes 1¼" to 2"	2½" to 6"
Low-pressure/ temp	251–250	1.5	1.5	2.0
Low temperature	120–200	1.0	1.0	1.5
Chilled water	40–55	0.5	0.75	1.0
Refrigerant or brine	below 40	1.0	1.5	1.5

R VALUES FOR FLOORS

Wall R-Value	Application	Description	Thickness (inches)	R Value Achieved
R-11	On concrete floor of dry waterproofed basement	Install pressure treated 2" × 4" sleepers 24" o.c. Lay R-11 fiberglass floor insulation batts between sleepers, apply vapor barrier and min. 5⁄8" plywood flooring	3½	11.8
R-12 to 14		As above, installing 3 1" layers fiberglass perimeter insulation or 1¾" polyisocyanurate, vapor barrier and 5⁄8" plywood	3 1¾	12.7 13.4
		As above installing 2" polyisocyanurate foam board, vapor barrier and 5⁄8" plywood	2	14.4
R-30 R-19 R-13	Floors over unheated crawl space or basement floors over unheated basements	Install glass fiber or rock wool batts, vapor barrier side up, and retain with "tiger teeth" (stiff wire rods pointed at each end ½" longer than joist spacing)	9½ 6½ 3⅝	30.0 19.0 13.0
Slab Perimeter R-8 to 10	Around exterior perimeter monolithic slab	Double layer 1" glass fiber perimeter insulation, hard-faced and top flashed	2	8.6
	Install 20"-wide insulation flat on fill inside perimeter of slab with 4" vertical at slab edge	Glass fiber perimeter insulation 2 layers *or* Extruded polystyrene foam insulation	2 2 2	8.6 8.0 10.0

SELECTED R-VALUES FOR BASEMENT/CRAWL SPACE WALLS

R-Value Desired	Application	Description	Insulation Type	Insulation Layers	Thickness Each (inches)	R-Value Achieved
R-25 to 30	Poured concrete or concrete block	Interior 2" × 4" × 24" o.c. wall offset 2⅝" *plus* the insulation types shown in next column.	Glass fiber or rock wool batt or blanket.	1 layer	6¼	29
			Polyisocyanurate foamboard on studs, faced with gypsum board.	1 layer	1½	29.8
			Glass fiberboard on exterior protected to grade or on interior faced with gypsum board.	1 layer	1½	29.8
			Polystyrene foamboard on exterior, protected to grade or on interior faced with gypsum board.	1 layer	2	27
	All-weather wood foundation	2" × 8" pressure-treated fame foundation	Fiberglass or rock wool batt or blanket faced with gypsum board.	1 layer	9½	25.7

		2" × 12" pressure-treated frame foundation	Fiberglass or rock wool batt or blanket faced with gypsum board.	1 layer	9½	30
R-20 to 21	All-weather wood foundation	2" × 8" pressure-treated wood foundation	Fiberglass or rock wool batt or blanket faced with gypsum board.	1 layer	6¼	19
		2" × 12" pressure-treated wood foundation	Fiberglass or rock wool batt or blanket faced with gypsum board.	1 layer	6¼	19
	Poured concrete or concrete block	Interior 2" × 4" × 24" o.c. wall offset 2⅝"	Fiberglass or rock wool batt or blanket faced with gypsum board.	1 layer	6¼	19
R-12 to 15	All-weather wood foundation	2" × 8" or 2" × 12" pressure-treated wood foundation	Fiberglass or rock wool batt or blanket faced with gypsum board.	1 layer 1 layer	3½ 3⅝	11 13
	Poured concrete or concrete block	Interior 2" × 4" × 24" o.c. wall	Fiberglass or rock wool batt or blanket faced with gypsum board.	1 layer 1 layer	3½ 3⅝	11 13
			Polyisocyanurate foamboard, gypsum faced.	1 layer	1½	10.8

BLOWN-IN INSULATION PROPERTIES

Requirements to Achieve Desired R Values					
Location	Insulation Type	Approx. Product Thickness Inches	R Value	Bags Per 1000 Sq. Ft.	Coverage Per Bag In Sq. Ft.
Attic	Cellulose, loose fiber	3.7	R-13	62	16.0
		5.2	R-19	86	11.5
		6.7	R-24	110	9.0
		8.7	R-32	154	6.5
		11.0	R-40	182	5.5
Walls	Cellulose, loose fiber	3.5	R-13	61	16.5
Attic	Rock wool, loose fiber	4.5	R-13	Variable	Variable
		6.6	R-19	Variable	Variable
		11.0	R-32	Variable	Variable
		13.7	R-40	Variable	Variable
Walls	Fiberglass, cubed	3.63	R-12	14.3	70
		4	R-13	15.8	63

Attic	Fiberglass, cubed	5.25	R-11	7.2	140
		8.75	R-19	12.4	81
		10.25	R-22	14.3	70
		12.25	R-26	16.9	59
		14.00	R-30	19.5	51
		17.75	R-38	24.8	40
Walls	Fiberglass, loose fiber	3.75	R-11	12	77
		6.50	R-19	22	45
		7.50	R-22	25	40
		8.75	R-26	29	34
		10.25	R-30	34	29
		12.75	R-38	43	23
Floored Attic	Fiberglass, cubed 2" × 6"	5.5	R-18	21.7	46
	2" × 8"	7.25	R-24	28.6	35
	2" × 10"	9.25	R-30	36.6	27

FLEXIBLE INSULATION PROPERTIES

	Fiberglass Batts, Blankets, 2 Rolls	Fiberglass Cubed and Loose Blowing Wool	Rock or Slag Wool Batts and Blanket	Rock Wool, Loose Blowing Wool
Used for:				
Attics	•	•	•	•
Sloping roofs	•		•	
Walls (new const.)	•		•	
Walls (existing)		•		•
Floors over unheated spaces	•		•	
R-Value/Inch	3.1	Attic 2.2–3.0 Walls 3.4–4.0	3.16	Attic 2.6–2.8 Walls 3.75
Widths and Lengths	15 or 23 × 94 16 or 24 × 48 Rolls: 25' R-25 39' R-19 70.5' R-11	NA	15 or 23 × 48 or 94	NA

Facings and Standard R-values	Unfaced, FS-25, foil & kraft 11, 13, 19, 22, 25, 30, 38. For R-values 60 and above, use multiple layers.	NA Blown to any R-value desired. See data on packaging to find minimum number of inches, number of lbs. per square foot, and min. number of bags per 1,000 sq. ft. necessary to achieve R-value.	FS-25, foil & kraft 11, 13, 19, 22, 30 (sold through contractors not readily available for D.I.Y. use.)	NA Blown to R-value desired. If exceeding 1.3 psf on 24" o.c. joists, gypsum manufacturers recommend thicker than ½" gypsum board be used for ceiling to prevent sagging due to weight.
Fire-resistance	Non-combustible			
Moisture absorption	Negligible	Negligible	Negligible	Negligible
Corrosion	None	None	None	None
Fungal-bacterial growth	Does not promote growth	Does not promote growth	Does not support growth	Does not support growth
Other	Self-supporting in stud space.		Easy to cut and install	Heavier

DUCT WRAP INSULATION

Type	Thickness	R-Value, Installed	R-Value, Packaged
Unfaced			
50	1½"	3.6	4.8
	2"	4.8	6.5
	3"	7.3	9.7
75	1"	2.6	3.4
	1½"	3.9	5.2
	2"	5.2	6.9
	3"	7.8	10.3
100	1"	2.8	3.7
	1½"	4.2	5.6
	2"	5.6	7.4
150	1"	3.0	4.0
	1½"	4.5	6.5
Faced			
75	1½"	3.8	5.0
	2"	5.0	6.7
	2½"	6.3	8.3
	3"	7.5	10.0

CAULKING POINTS

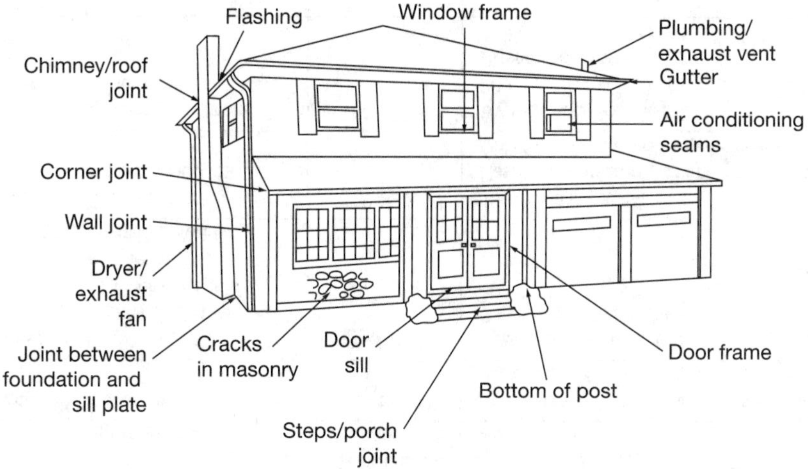

Flashing
Window frame
Plumbing/
exhaust vent
Gutter
Chimney/roof
joint
Air conditioning
seams
Corner joint
Wall joint
Dryer/
exhaust
fan
Joint between
foundation and
sill plate
Cracks
in masonry
Door
sill
Door frame
Bottom of post
Steps/porch
joint

CAULK APPLICATIONS

Caulk	Application	Temperature Range	Notes
Acrylic Latex	Indoor cracks.	−20 to 180°F	Dries quickly. Can be painted with oil or latex. Avoid use outdoors or in wet locations.
Asphalt-based Caulks	Waterproofing chimneys, dormer flashings, skylights, roof seams, bonding polyethylene.	40 to 150°F	Waterproof, good bonding. Check estimated resilient life.
Butyl Sealant	Driveway cracks, gutters, windows.	−20 to 200°F	10—20 year life. Resistant to heat, oil, chemicals and water. Only for openings no wider or deeper than ¼".
Ethylene Elastomeric Copolymer	Exterior applications, all surfaces.	−30 to 180°F	UV resistant. Good elasticity. 20 to 25 year life. Plastics need priming. 24 hour cure. Paintable.

Neoprene	Metal and mortar joints, flashing and seams. Driveway sealing.	−40 to 250°F	15 to 20 year life Resistant to heat, oil, and chemicals. Adheres to any surface. Resistance to movement. Chemical resistant. Dries in 4 hours.
Nitrile	Outdoor and for metal, wood, plastics, glass, ceramics, plaster, drywall, and concrete.	40 to 210°F	Long life. Excellent for narrow cracks and joints. Resists movement and vibration.
Oil-Based	Holes in metal, wood, or masonry. Indoors.	–20 to 180°F	Inexpensive. 1-year to 3-year life.
Polybutane	Indoors or outdoors for openings, cracks, gutters, windows and doors.	–20 to 200°F	Resistant to heat, oil, and chemicals. Good water resistance. 10 to 20 year life. Use only on openings ¼" or less, wide or deep. Dries slowly.
Polysulphides	Between glass and frame. High movement joints.	−40 to 250°F	Excellent flexibility, stretch, and bonding. Life expectancy 20 years. Must be applied to dry surfaces.

CAULK APPLICATIONS *(cont.)*

Caulk	Application	Temperature Range	Notes
Polyurethane	Around vents, ducts, and pipes. Sealing sill plates, top plates, T-joints, and wood-to-wood joints. Filling deep voids.	−40 to 120°F	Can be trimmed within 10 minutes, cures in 8 hours. Approximately R-5 per inch of depth and stops drafts. May turn brown. Requires application skill to avoid waste. Keep out of the reach of children. Combustible; use only in well-ventilated area. Moisten dry or porous surfaces before application.
Rubber Caulk	Around windows and doors.	0 to 100°F	Does not collect dirt. Can be painted.
	Gutters and down spouts.	0 to 100°F	
Silicone	Almost any use.	−75 to 450°F	Seals, bonds, and insulates. Top performer. Will not bond to tetrachlorethylene. Does not crack or become brittle with age. Does not sag, slump, run, or drip. Good dielectric properties. Work time 3–10 minutes, vulcanizes in 24 hours. Cure 4–7 days. Requires moist air to cure. Most types will not accept paint. Solvents: naptha, xylene, or toluene.

Siliconized Acrylic Latex	Indoor and outdoor cracks.	–20 to 180°F	Easy to use. Dries fast. Good flexibility.
Solvent-Based Acrylics	Adhesive and sealant. Caulking around windows, doors, sinks, bathtubs and showers. Holes and cracks in wood, plaster and drywall.	–40 to 300°F	Can be painted. Mildew resistant. Adheres to ceramic tiles. Some types require preheating before use. Does not adhere to Teflon, polyethylene, silicone, or wet surfaces. Difficult to apply.
Thermoplastic Caulking Sticks	General use, tight bonded seal.	0 to 350°F	No mess, sets quickly and adheres firmly.
Urethane	General use. Adheres to almost any surface except some plastics.	0 to 250°F	Tough, abrasion-resistant. Can be used in high-movement joints.
Vinyl Caulk	All-purpose adhesive caulk. Bonds to wood, masonry, ceramics, glass, and metal.	20 to 180°F	Resilient, mildew-resistant, water-resistant, paintable. Nontoxic, nonflammable. Water cleanup.
Vinyl Foam Tape	Press into cracks.	20 to 180°F	(Same as above.)

WEATHER-STRIPPING DIAGRAMS

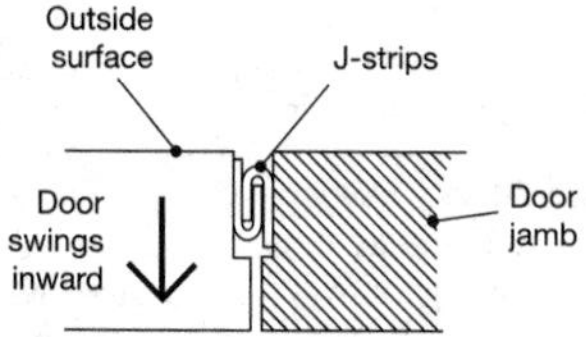

Metal J Strips

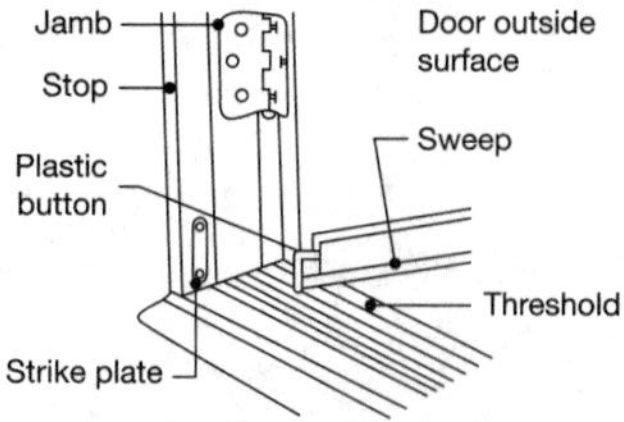

Automatic Sweep

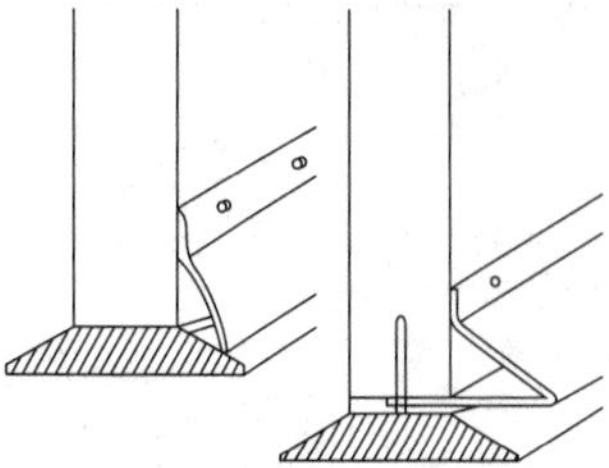

Metal Drip Caps

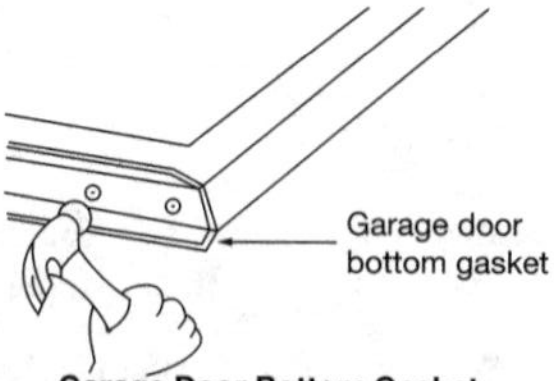

Garage Door Bottom Gasket

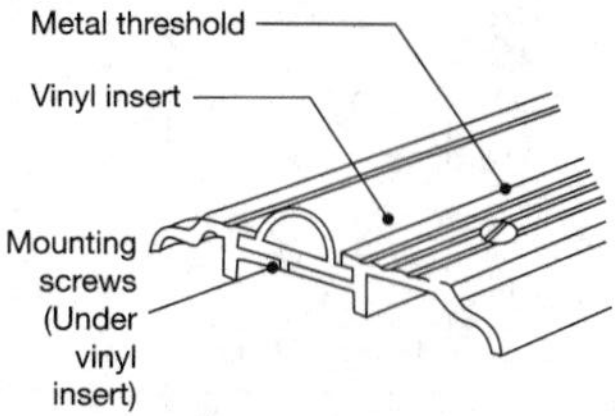

Bulb Threshold

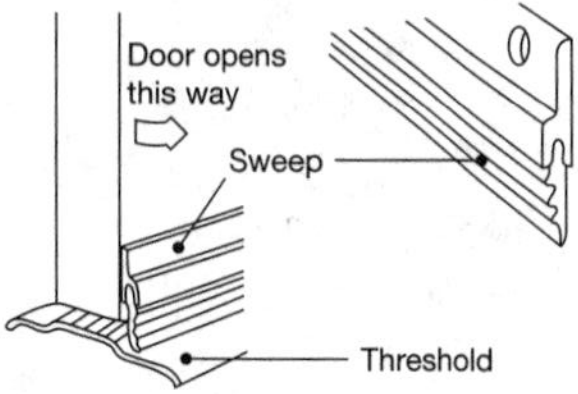

Storm Door Sweep

CHAPTER 5
Painting and Finishing

DRYING TIMES FOR PRIMERS	
Primer	**Time (hours)**
Acrylic latex block filler	2 to 4
Alkyd flat enamel	2 to 4
Alkyd metal primer	4 to 5
Alkyd primer	2 to 4
Alkali-resistant enamel	2 to 4
Epoxy	5 to 8
Lacquer sealer	1
Latex flat	2 to 4
Latex primer	2 to 4
Liquid filler	24
Non-grain-raising (NGR)	3 to 4
Oil-base	24
Paste wood filler	24
Polyurethane	12
Portland cement masonry primer	24
Primer sealer	¾ to 1
Sanding sealer	1
Shellac	3 to 4
Stain blocking primer	24
Stains	24
Varnish	24 to 36
Water-base	24
Zinc metal primer	24

DRYING TIMES OF COATINGS			
Material	**Touch**	**Recoat**	**Rub**
Lacquer	1–10 min.	1½–3 hrs.	16–24 hrs.
Lacquer sealer	1–10 min.	30–45 min.	1 hr. (sand)
Paste wood filler	—	24–48 hrs.	—
Paste wood filler (Q.D.)	—	3–4 hrs.	—
Water stain	1 hr.	12 hrs.	—
Oil stain	1 hr.	24 hrs.	—
Spirit stain	Zero	10 min.	—
Shading stain	Zero	Zero	—
Non-grain raising stain	15 min.	3 hrs.	—
NGR stain (quick-dry)	2 min.	15 min.	—
Pigment oil stain	1 hr.	12 hrs.	—
Pigment oil stain (Q.D.)	1 hr.	3 hrs.	—
Shellac	15 min.	2 hrs.	12–18 hrs.
Shellac (wash coat)	2 min.	30 min.	—
Varnish	1½ hrs.	18–24 hrs.	24–48 hrs.
Varnish (Q.D. synthetic)	½ hr.	4 hrs.	12–24 hrs.
Note: Average times. Different products will vary.			

BRISTLE APPLICATIONS

Type of Paint	Type of Bristle
Oil paint and oil stain	Natural bristles or polyester
Alkyd, epox, enamel, varnish, lacquer	Polyester
Latex, water-based paint, water-based stains	Polyester, nylon, and polypropylene
Adhesive and paste	Stiff vegetable fiber brushes or trowels

PAINTBRUSH APPLICATIONS

Application	Brush Type	Width (inches)	Length of Bristle (inches)
Ceiling	Square	4	3¼–4
Doors and shelves	Chiseled edge	2, 2½, 3	2½–3
Enamel and varnish	Chiseled edge	1, 1½	2–2¼
Floors	Square	4	3¼–4
Furniture	Square	1, 1½, 2, 2½	2¼–3
Masonry Walls	Square	6	3¼–4
Moldings	Chiseled or angular	1½	2¼–2¾
Sashes (round or flat)	Round square	1	2¼–2¾
	Flat beveled	1½	2–2¼
Siding	Square	4, 5	3¼–4
Trim	Chiseled or beveled	1	2¼
Walls (interior)	Square	4, 5	3¼–4
Walls (exterior)	Square	6	3¼–4
Window frames	Square	2	2¼–2¾
Wrought iron metal furniture	Square	1, 1½	2¼–2¾

APPROXIMATE PAINT REQUIREMENTS FOR INTERIORS AND EXTERIORS

Distance around the Room	Ceiling Height 8 Feet	Ceiling Height 8½ Feet	Ceiling Height 9 Feet	Ceiling Height 9½ Feet	Paint for Ceiling	Finish for Floors	For Each Door or Window
30 feet	⅝ gallon	⅝ gallon	¼ gallon	¾ gallon	1 pint	1 pint	
35 feet	¾ gallon	¾ gallon	¼ gallon	⅞ gallon	1 quart	1 pint	Each window and frame requires ¼ pint
40 feet	⅞ gallon	⅞ gallon	⅞ gallon	1 gallon	1 quart	1 quart	
45 feet	⅞ gallon	1 gallon	1 gallon	1⅛ gallons	3 pints	1 quart	
50 feet	1 gallon	1⅛ gallons	1⅛ gallons	1¼ gallons	3 pints	1 quart	
55 feet	1⅛ gallons	1⅛ gallons	1¼ gallons	1¼ gallons	2 quarts	3 pints	Each door and frame requires ½ pint
60 feet	1¼ gallons	1¼ gallons	1⅜ gallons	1⅜ gallons	2 quarts	3 pints	
70 feet	1⅜ gallons	1½ gallons	1½ gallons	1⅝ gallons	3 quarts	2 quarts	
80 feet	1½ gallons	1⅝ gallons	1¼ gallons	1⅞ gallons	1 gallon	5 pints	

APPROXIMATE PAINT REQUIREMENTS FOR INTERIORS AND EXTERIORS *(cont.)*

Distance around the House	Average Height 12 Feet	Average Height 15 Feet	Average Height 18 Feet	Average Height 21 Feet	Average Height 24 Feet
60 feet	1 gallon	1¼ gallons	1½ gallons	1¼ gallons	2 gallons
76 feet	1¼ gallons	1½ gallons	2 gallons	2¼ gallons	2½ gallons
92 feet	1½ gallons	2 gallons	2½ gallons	2¾ gallons	3 gallons
108 feet	1¾ gallons	2¼ gallons	2¾ gallons	3¼ gallons	3¾ gallons
124 feet	2 gallons	2½ gallons	3¼ gallons	3¾ gallons	4¼ gallons
140 feet	2½ gallons	3 gallons	3½ gallons	4 gallons	4½ gallons
156 feet	2¾ gallons	3¼ gallons	4 gallons	4½ gallons	5¼ gallons
172 feet	3 gallons	3¼ gallons	4½ gallons	5 gallons	5¼ gallons

On interior work, for rough, sand-finished walls or unpainted gypsum board, add 50% to quantities; for each door or window, deduct ½ pint of materials for walls. For trim, add ⅛ to ⅕ of the amount required for the body. For exterior blinds, ½ gallon will cover 12 to 14 blinds, one coat.

EXTERIOR PAINT SELECTION CHART

	Aluminum	Cement Base Paint	Exterior Clear Finish	House Paint	Metal Roof Paint	Porch-and-Deck Paint	Primer or Undercoater	Rubber Base Paint	Spar Varnish	Transparent Sealer	Trim-and-Trellis Paint	Wood Stain	Metal Primer
Wood													
Natural finish	—	—	P	—	—	—	—	—	P	—	—	P	—
Porch floor	—	—	—	—	—	P	—	—	—	—	—	—	—
Shingle roof	—	—	—	—	—	—	—	—	—	—	—	—	—
Shutters and trim	—	—	—	P+	—	—	P	—	—	—	P+	—	—
Siding	P	—	—	P+	—	—	P	—	—	—	—	—	—
Windows	P	—	—	P+	—	—	P	—	—	—	P+	—	—
Masonry													
Asbestos cement	—	—	—	P+	—	—	P	P	—	—	—	—	—
Brick	P	P	—	P+	—	—	P	P	—	P	—	—	—
Cement & cinder block	P	P	—	P+	—	—	P	P	—	P	—	—	—
Cement porch floor	—	—	—	—	—	P	—	P	—	—	—	—	—
Stucco	P	P	—	P+	—	—	P	P	—	P	—	—	—
Metal													
Copper	—	—	—	—	—	—	—	—	P	—	—	—	—
Galvanized	P+	—	—	P+	—	—	P	—	P	—	P+	—	P
Iron	P+	—	—	P+	—	—	—	—	—	—	P+	—	P
Roofing	—	—	—	—	P+	—	—	—	—	—	—	—	P
Siding	P+	—	—	P+	—	—	—	—	—	—	P+	—	P
Windows, aluminum	P	—	—	P+	—	—	—	—	—	—	P+	—	P
Windows, steel	P+	—	—	P+	—	—	—	—	—	—	P+	—	P

"P" indicates preferred coating for this surface. "P+" indicates that a primer or sealer may be necessary before the finishing coat or coats (unless the surface has been previously finished).

INTERIOR FINISH SELECTION

Surfaces

Interior Primers	Acoustical Tile	Aluminum	Brick	Cement Blocks	Concrete	Concrete Floors	Drywall	Galvanized Steel	Iron and Steel	Paneling	Plaster	Wood Flooring	Wood Trim	Finish Coats
Acrylic Latex Block Filter				●										Latex masonry paint or enamel
Alkali Resistant Enamel				●	●									Alkali-resistant enamel or latex paint
Alkyd Flat Enamel										●			●	Latex flat or alkyd semigloss or gloss enamel
Alkyd Metal Primer		●							●					Latex flat or alkyd semigloss or gloss enamel
Alkyd Primer	●				●						●			Only use latex flat paint on acoustical tile
Clear Wood Sealer												●	●	Any latex or alkyd paint or polyurethane over stain
Enamel Undercoat										●			●	Any latex or alkyd paint
Exterior Masonry Alkyd Paint			●	●										Latex or alkyd masonry paint
Latex Flat Wall Paint	●						●				●			Latex flat wall paint
Latex Primer	●						●				●			Any latex or alkyd paint or enamel (only use latex flat on acoustical tile)
Latex Metal Primer		●						●	●					Any latex paint or enamel
Portland Cement Masonry Paint				●	●	●								Portland cement masonry paint
Portland Cement Metal Primer								●						Portland cement metal paint
Surface Conditioner			●		●							●		Latex or alkyd semigloss or gloss enamel
Use Topcoat as Primer						●						●		Same as primer: polyurethane, clear finish, epoxy enamel, moisture-cured urethane, 2-part epoxy, or polyurethane
Wood Filler										●		●	●	Clear wood sealer
Zinc Metal Primer								●						Any alkyd paint or enamel

PRIMERS SELECTION

Types	Aluminum	Aluminum Gutters, Downspouts, or Siding	Asphalt Roofing	Brick, Block, Concrete, Stone, or Stucco	Brass, Bronze, or Copper	Galvanized Steel	Iron or Steel (bright)	Metal Roofing	Plywood	Porches and Decks	Pressure-Treated Lumber	Shakes, Shingles, Wood Clapboard, or Siding	Terne Roofing	Wood Trim	Finish Coat
Alkyd-base Metal Primer	●	●			●	●	●								Alkyd house paint, alkyd trim enamel, aluminum paint, alkyd-base metal exterior enamel
Alkyd Deck Primer										●	●				Alkyd deck paint, (2-part epoxy, 2-part polyurethane)
Alkyd primer														●	Alkyd house paint or alkyd trim enamel
Alkali-Resistant Coating				●											Alkali-resistant coating etc.
Asphalt Roof Paint			●												Aluminum or bituminous roof coating
Iron Oxide Primer													●		Alkyd, oil-based latex house or trim or exterior metal enamel
Latex Metal Primer					●	●	●								Latex house and trim paint
Latex Primer												●		●	Latex house paint, latex trim enamel alkyd or oil-based paints
Masonry Surface Conditioner				●											Oil-base alkyd latex house, trim, or masonry paint; alkali-resistant coating

PRIMERS SELECTION *(cont.)*

Types	Aluminum	Aluminum Gutters, Downspouts, or Siding	Asphalt Roofing	Brick, Block, Concrete, Stone, or Stucco	Brass, Bronze, or Copper	Galvanized Steel	Iron or Steel (bright)	Metal Roofing	Plywood	Porches and Decks	Pressure-Treated Lumber	Shakes, Shingles, Wood Clapboard, or Siding	Terne Roofing	Wood Trim	Finish Coat
	Surfaces														
Oil-base Metal Primer	●	●			●	●	●								Oil-base exterior metal paint; oil-base house and trim paint
Oil Primer										●	●	●			Oil-base flat, semigloss, or gloss house paint barn paint
Portland Cement Masonry Paint				●											Portland cement masonry paint
Portland Cement Metal Paint						●	●								Portland cement paint for metal
*Stain-sealing Primer									●		●	●		●	Any alkyd, latex or oil-base house or trim for plywood use on acrylic-latex
Use Topcoat as Primer			●												Aluminum or bituminous roof coating
Water Repellent Preservative									●						Stain (semi-transparent or solid color)
Zinc Chromate Primer						●	●	●							Aluminum paint, bituminous roof coating
Zinc-dust Primer						●	●								Aluminum paint, bituminous roof coating

COMPARISON OF PAINT BINDERS' PRINCIPAL PROPERTIES

	Alkyd	Cement	Epoxy	Latex	Oil	Phenolic	Rubber	Moisture Curing Urethane	Vinyl
Ready for use	Yes	No	No[c]	Yes	Yes	Yes	Yes	Yes	Yes
Brushability	A	A	A	+	+	A	A	A	–
Odor	+[a]	+	–	+	A	A	A	–	–
Cure normal temp.	A	A	A	+	–	A	+	+	+
Cure low temp.	A	A	–	–	–	A	+	+	+
Film build/coat	A	+	+	A	+	A	A	+	–
Safety	A	+	–	+	A	A	A	–	–
Use on wood	A	–	A	A	A	A	–	A	–
Use on fresh conc.	–	+	+	+	–	–	+	A	+
Use on metal	+	–	+	–	+	+	A	A	+
Corrosive service	A	–	+	–	–	A	A	A	+
Gloss – choice	+	–	+	–	A	+	+	A	A
Gloss – retention	+	X	–	X	–	+	A	A	+
Color – initial	+	A	A	+	A	–	+	+	+
Color – retention	+	–	A	+	A	–	A	–	+
Hardness	A	+	+	A	–	+	+	+	A
Adhesion	A	–	+	A	+	A	A	+	–
Flexibility	A	–	+	+	+	A	A	+	+
Resistance to:									
Abrasion	A	A	+	A	–	+	A	+	+
Water	A	A	A	A	A	+	+	+	+
Acid	A	–	A	A	–	+	+	+	+
Alkali	A	+	+	A	–	A	+	+	+
Strong solvent	–	+	+	A	–	A	–	+	A
Heat	A	A	A	A	A	A	+[b]	A	–
Moisture permeability	Mod.	V. High	Low	High	Mod.	Low	Low	Low	Low

+ = Among the best for this property   – = Among the poorest for this property   A = Average   X = Not applicable   [a] Odorless type   [b] Special types   [c] Two component type

SINGLE-ROLL WALLPAPER REQUIREMENTS

Size of Room	Height of Ceiling			Yards of Border	Rolls for Ceiling
	8'	9'	10'		
4 × 8	6	7	8	9	2
4 × 10	7	8	9	11	2
4 × 12	8	9	10	12	2
6 × 10	8	9	10	12	2
6 × 12	9	10	11	13	3
8 × 12	10	11	13	15	4
8 × 14	11	12	14	16	4
10 × 14	12	14	15	18	5
10 × 16	13	15	16	19	6
12 × 16	14	16	17	20	7
12 × 18	15	17	19	22	8
14 × 18	16	18	20	23	8
14 × 22	18	20	22	26	10
15 × 16	15	17	19	23	8
15 × 18	16	18	20	24	9
15 × 20	17	20	22	25	10
15 × 23	19	21	23	28	11
16 × 18	17	19	21	25	10
16 × 20	18	20	22	26	10
16 × 22	19	21	23	28	11
16 × 24	20	22	25	29	12
16 × 26	21	23	26	31	13
17 × 22	19	22	24	23	12
17 × 25	21	23	26	31	13
17 × 28	22	25	28	32	15
17 × 32	24	27	30	35	17
17 × 35	26	29	32	37	18
18 × 22	20	22	25	29	12
18 × 25	21	24	27	31	14
18 × 28	23	26	28	33	16

This chart assumes use of the standard roll of wallpaper, 8 yards long and 18 inches wide. Deduct one roll of side wallpaper for every two doors or windows of ordinary dimensions, or for each 50 square feet of opening.

FLOOR FINISH MAINTENANCE		
Flooring	**Finish**	**Cleaning**
Resilient flooring: asphalt tile, linoleum, rubber vinyl, and terrazzo	Liquid or paste floor wax with Carnuba base wax, clear acrylic or urethane.	Damp mop with soap and water.
Wood floors and lacquer, retardant spray varnish, enamel, or polyurethane finish	Paste or spray wax.	Silicone and wipe furniture as directed.
Wood treated with penetrating resin	Second coat of resin.	Soap and water.
Wood with shellac and French polish	Use paste or liquid wax and silicone spray wax.	Oiled cloth. No water or alcohol.
Wood finished with rubbed oil	No wax. Use a linseed-oiled cloth and rub in.	3 parts turpentine to 1 part linseed oil.

CHAPTER 6
Plumbing

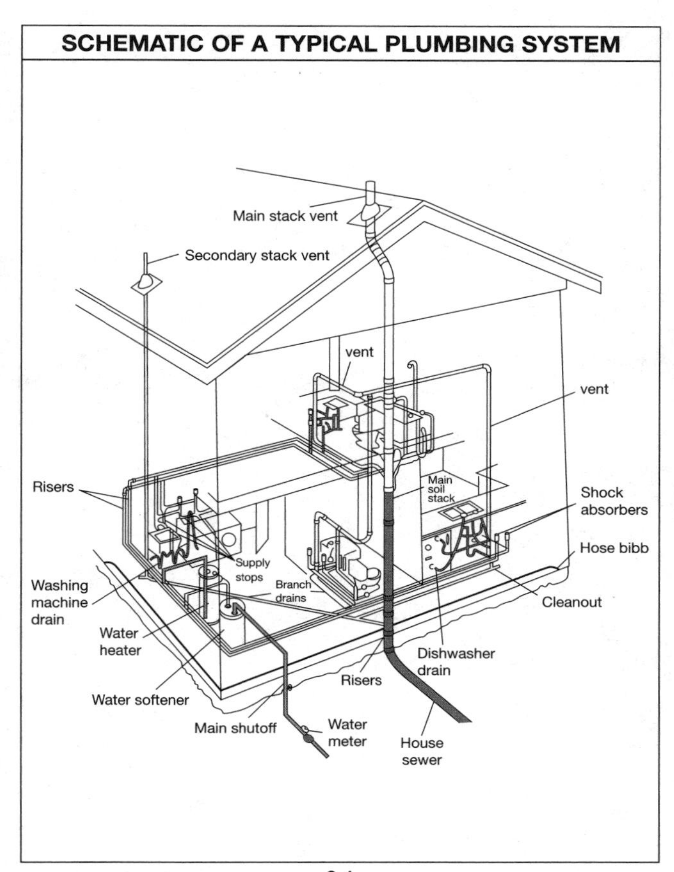

FITTING DIAGRAMS

1/16 Bend

1/8 Bend

1/6 Bend

1/4 Bend

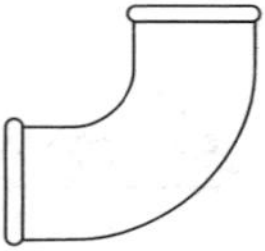

Long 1/8 Bend

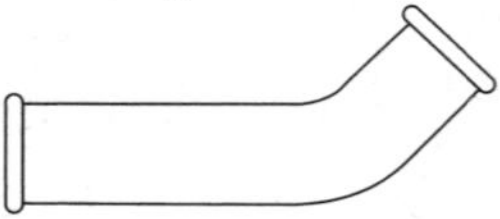

Long Sweep

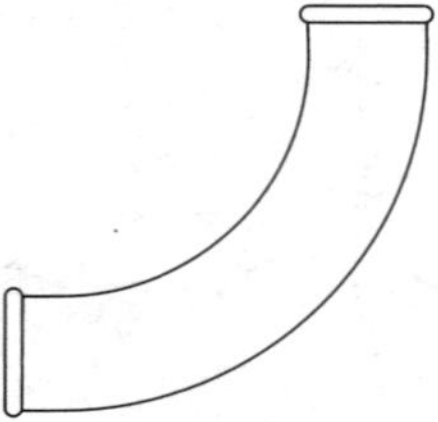

Sanitary Tee

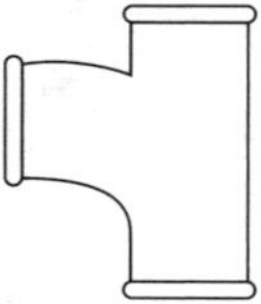

Y

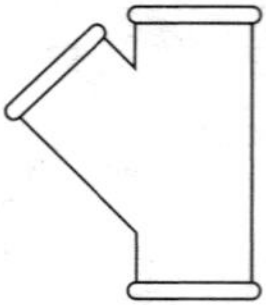

Combination Tee

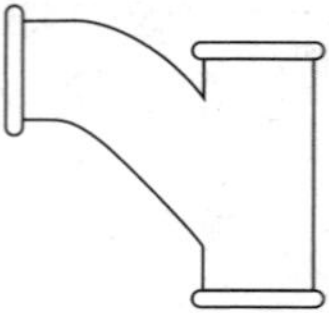

FITTING DIAGRAMS *(cont.)*

Tap Short
1/4 Bend

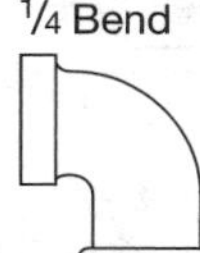

Tap Sanitary Tee

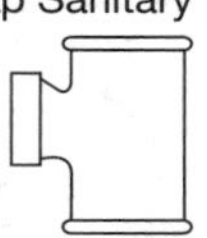

Tap Extension Piece

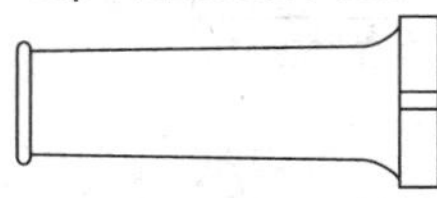

Tap Adapter

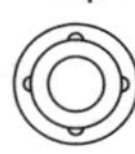

Closet Flange
(Slot & Notch)

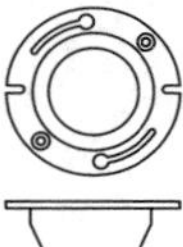

Hub Adapter

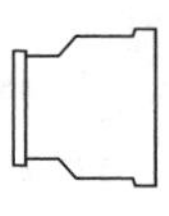

P Trap

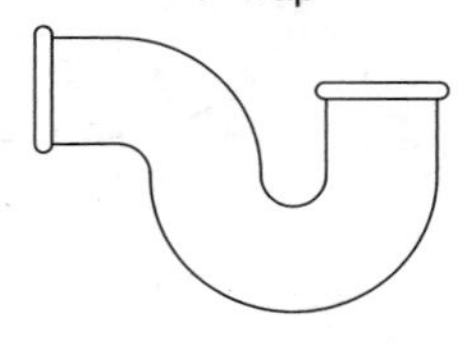

Short Reducer

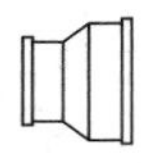

Test Tee Less Plug

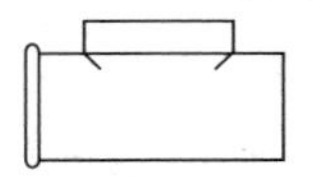

Blind Plug

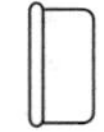

FITTING DIAGRAMS *(cont.)*

Sixteenth Bend

Short Eighth Bend

Long Eighth Bend

Cleanout Tee with BTS on Co-Opening

Stack Base Fitting

Closet Bend

FITTING DIAGRAMS *(cont.)*

Sixth Bend

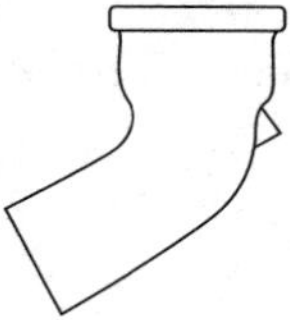

Quarter Bend

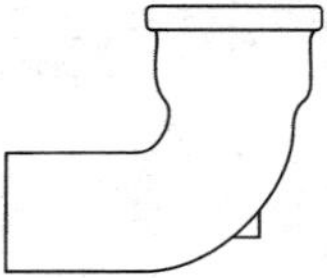

Long Sweep Bend

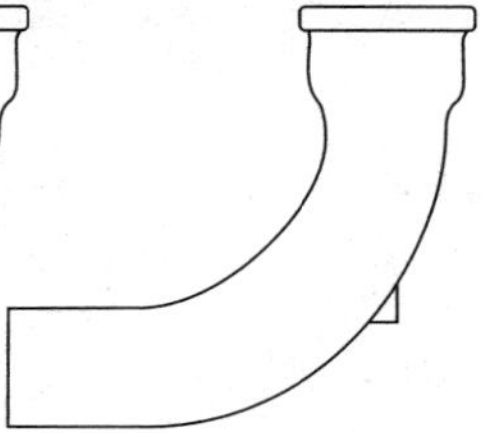

Y

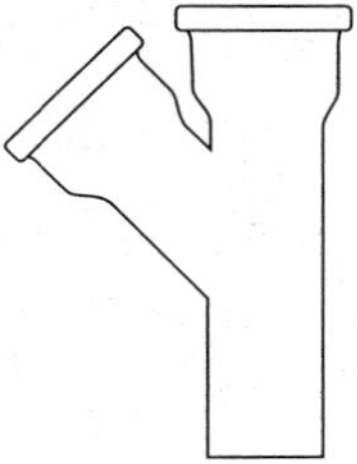

Combination Y and Eighth Bend

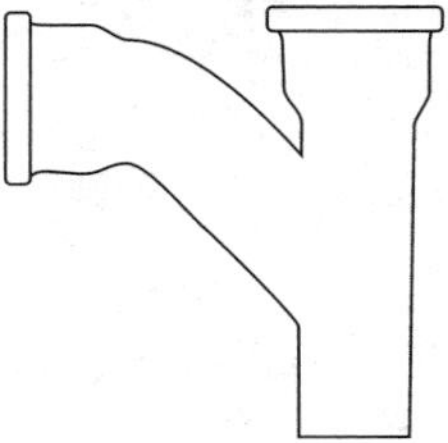

Sanitary Tee

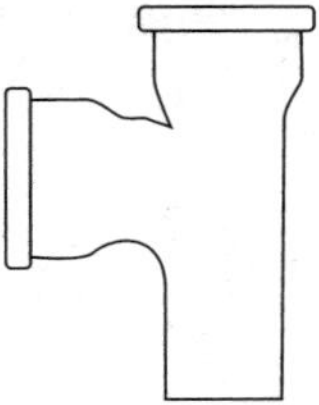

Tapped Tee

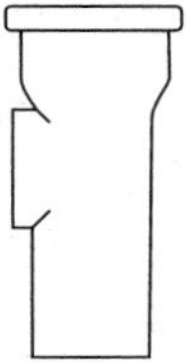

FITTING DIAGRAMS *(cont.)*

Cleanout Plug

Pipe Plug

Reducer

Adapter

Closet Flange

FITTING DIAGRAMS *(cont.)*

P Trap

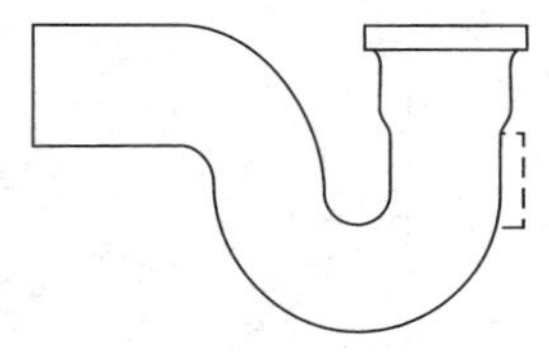

Floor Drain

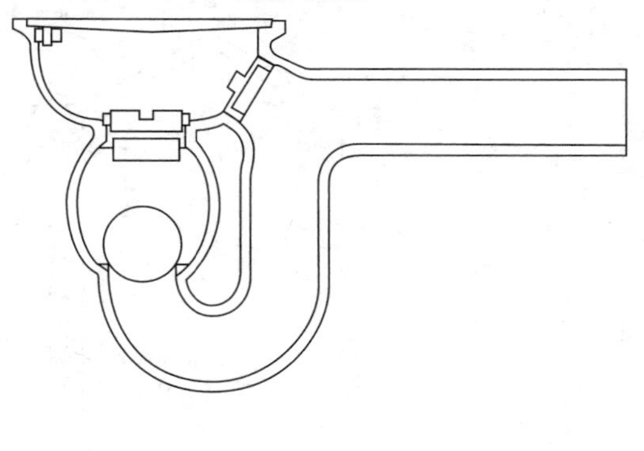

DRAIN AND VENT LINE DIAGRAMS

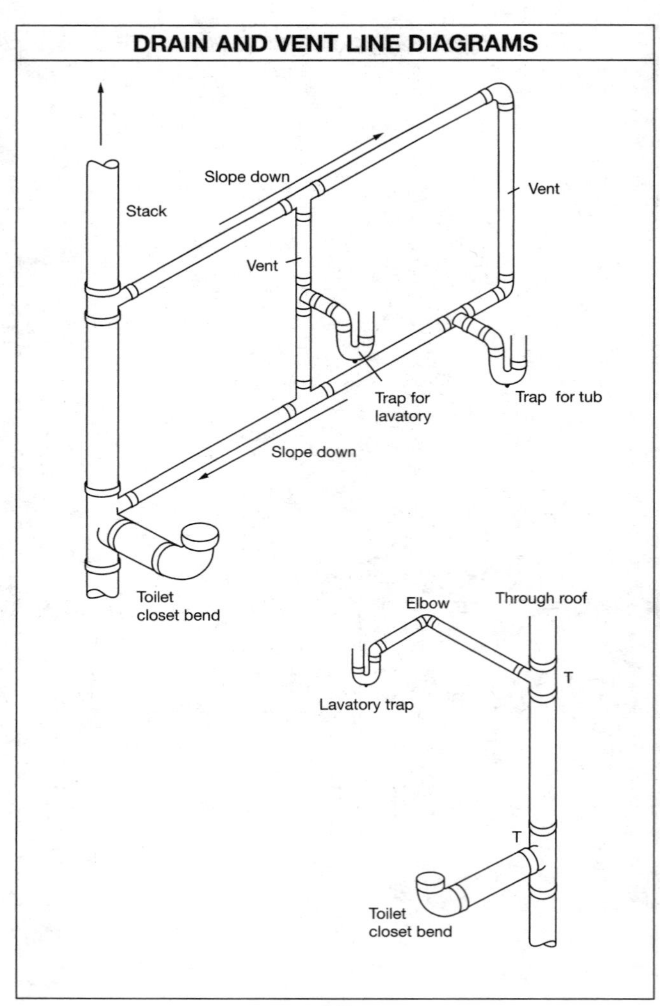

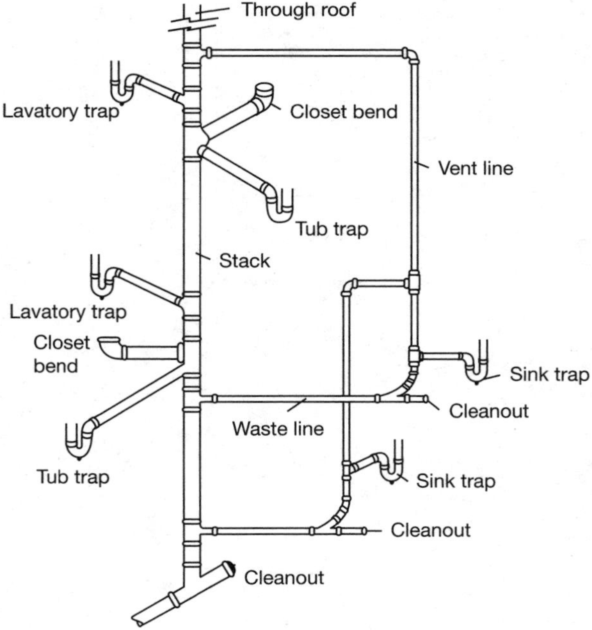

DRAIN AND VENT LINE DIAGRAMS *(cont.)*

Through roof

Lavatory trap

Closet bend

Vent line

Tub trap

Stack

Lavatory trap

Closet bend

Sink trap

Cleanout

Waste line

Tub trap

Sink trap

Cleanout

Cleanout

DRAIN AND VENT LINE DIAGRAMS *(cont.)*

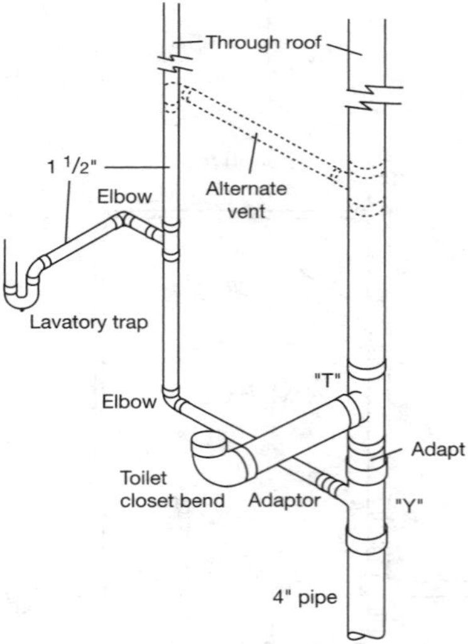

Through roof

1 1/2"

Elbow

Alternate vent

Lavatory trap

"T"

Elbow

Adapt

Toilet closet bend

Adaptor

"Y"

4" pipe

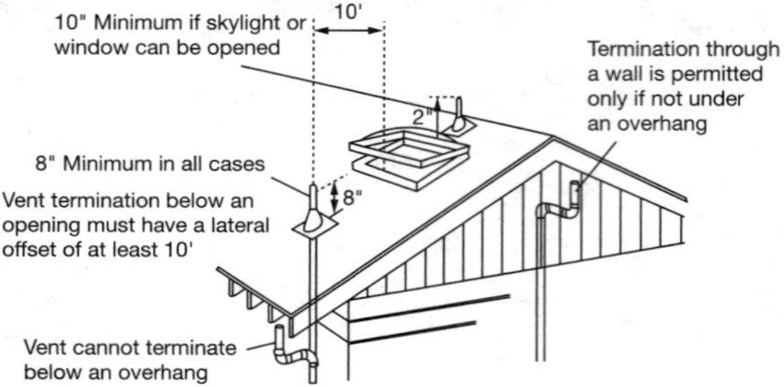

10" Minimum if skylight or window can be opened

10'

Termination through a wall is permitted only if not under an overhang

2"

8" Minimum in all cases

8"

Vent termination below an opening must have a lateral offset of at least 10'

Vent cannot terminate below an overhang

TRAPS AND VENTS

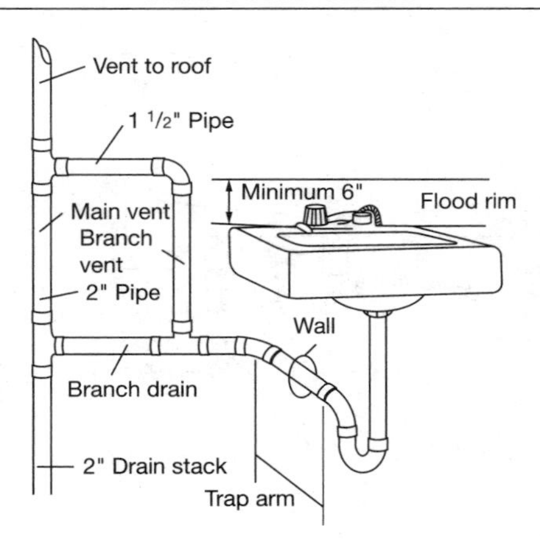

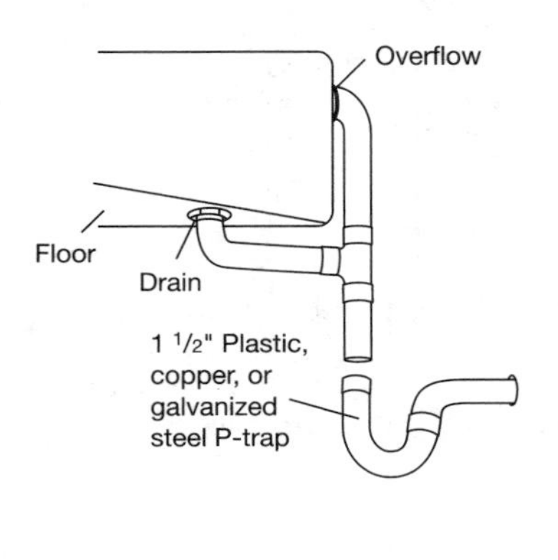

TRAPS AND VENTS *(cont.)*

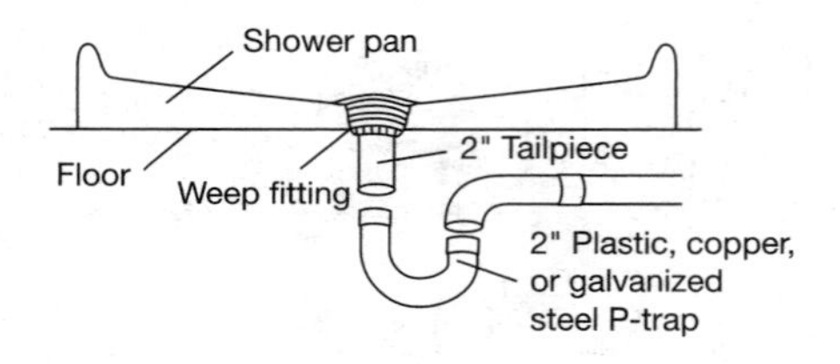

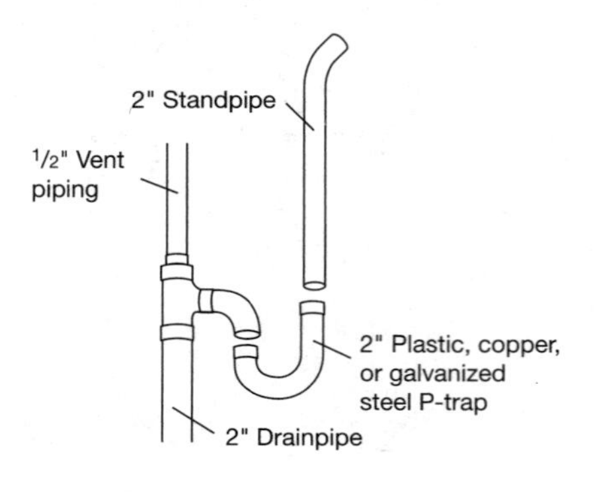

TRAPS AND VENTS *(cont.)*

PLASTIC PIPE

PVC and ABS are the most common types of plastic pipe. However, the correct primers and solvents must be used on each type or the joints will not seal properly and overall strength will be diminished.

Type	Characteristics and Guidelines
PVC	**Polyvinyl chloride, Type 1, Grade 1**. Strong, rigid, and resistant to a variety of caustic fluids. PVC is easy to work with and readily available. Maximum usable temperature is 140°F and pressure ratings start at a minimum of 125 to 200 psi. PVC can be used with water, gas, and drainage systems. DO NOT use with hot water systems.
ABS	**Acrylonitrile butadiene styrene, Type 1**. Like PVC, ABS is strong, rigid and resistant to a variety of caustic fluids and is very common, easy to work with, and readily available. Maximum usable temperature is 160°F at low pressure. ABS is generally used as a DWV (drain, waste, and vent) line.
CPVC	**Chlorinated polyvinyl chloride**. Similar to PVC but designed specifically for piping water at up to 180°F. CPVC has a pressure rating of 100 psi.
PE and **PEX**	**Polyethylene and polyethylene cross-linked**. Both are flexible pipes normally used for low-pressure water systems such as sprinklers.
PB	**Polybutylene**. A flexible pipe for pressurized water systems both hot and cold. Use only compression and banded type joints for installation.
Polypropylene	A lightweight material that is rated to 180°F at low pressure. Highly resistant to caustic fluids such as acids, bases, and solvents. Recommended for laboratory plumbing applications.
PVDF	**Polyvinylidene fluoride.** Strong, durable, and resistant to abrasion, acids, bases, and solvents that is rated to 280°F. Recommended for laboratory plumbing applications.
FRP Epoxy	A thermosetting plastic over fiberglass. Very strong with excellent caustic fluid resistance properties and is rated to 220°F. Also used in laboratory applications.

PVC PIPE SUPPORT SPACING IN FEET

Schedule 40

Nominal Pipe Size (in.)	Temperature Range (F)				
	60°	80°	100°	120°	140°
¼	3.75	3.50	3.00	2.50	2.00
½	4.25	4.00	3.50	3.00	2.50
¾	4.50	4.25	4.00	3.50	3.00
1	5.00	4.75	4.50	3.75	3.25
1¼	5.25	5.00	4.75	4.00	3.50
1½	5.50	5.25	5.00	4.25	3.75
2	6.00	5.50	5.00	4.50	4.00
2½	6.75	6.25	5.75	4.75	4.25
3	7.25	6.75	6.25	5.25	4.50
4	7.75	7.50	6.75	6.00	4.75
6	8.75	8.50	7.75	6.50	5.25
8	9.75	9.25	8.50	7.75	6.00
10	10.25	9.75	9.00	8.00	6.75
12	11.00	10.25	9.75	8.25	7.25

Schedule 80

Nominal Pipe Size (in.)	Temperature Range (F)				
	60°	80°	100°	120°	140°
¼	4.25	4.00	3.75	3.00	2.75
½	4.50	4.25	4.00	3.75	3.00
¾	4.75	4.50	4.25	4.00	3.50
1	5.00	4.75	4.50	4.25	3.75
1¼	6.00	5.00	4.75	4.50	4.25
1½	6.50	5.50	5.25	5.00	4.75
2	6.75	5.75	5.50	5.25	5.00
2½	7.25	6.75	5.75	5.50	5.25
3	7.75	7.50	7.00	6.25	5.50
4	9.00	8.75	7.25	6.50	5.75
6	9.75	9.50	8.50	7.75	6.50
8	11.00	10.25	9.75	8.75	7.00
10	11.50	10.50	10.25	9.50	7.75
12	12.50	12.25	11.50	10.25	8.75

STEEL PIPE DIMENSIONS AND WEIGHTS

Schedule 40

Nominal Pipe Size (in.)	Outside Diameter (in.)	Wall Thickness (in.)	Inside Diameter (in.)	Pipe Weight lbs. per ft.
1/8	.405	.068	.269	.245
1/4	.540	.088	.364	.425
3/8	.675	.091	.493	.568
1/2	.840	.109	.622	.851
3/4	1.050	.113	.824	1.131
1	1.315	.133	1.049	1.679
1 1/4	1.660	.140	1.380	2.273
1 1/2	1.900	.145	1.610	2.718
2	2.375	.154	2.067	3.653
2 1/2	2.875	.203	2.469	5.793
3	3.500	.216	3.068	7.580
3 1/2	4.000	.226	3.548	9.110
4	4.500	.237	4.026	10.790
5	5.563	.258	5.047	14.620
6	6.625	.280	6.065	18.970
8	8.625	.322	7.981	28.550
10	10.750	.365	10.020	40.480
12	12.750	.375	12.000	49.560
14	14.000	.500	13.000	72.090
16	16.000	.500	15.000	82.770
18	18.000	.562	16.876	104.750
20	20.000	.593	18.814	122.910

PIPE THREADING DIMENSIONS (NPT)

Nominal Pipe Size (in.)	Threads Per Inch	Approximate Length of Thread (in.)	Approximate Number of Threads To Be Cut	Approximate Total Thread Makeup, Hand and Wrench (in.)
1/8	27	3/8	10	1/4
1/4	18	5/8	11	3/8
3/8	18	5/8	11	3/8
1/2	14	3/4	10	7/16
3/4	14	3/4	10	1/2
1	11 1/2	7/8	10	9/16
1 1/4	11 1/2	1	11	9/16
1 1/2	11 1/2	1	11	9/16
2	11 1/2	1	11	5/8
2 1/2	8	1 1/2	12	7/8
3	8	1 1/2	12	1
3 1/2	8	1 5/8	13	1 1/16
4	8	1 5/8	13	1 1/16
5	8	1 3/4	14	1 3/16
6	8	1 3/4	14	1 3/16
8	8	1 7/8	15	1 5/16
10	8	2	16	1 1/2
12	8	2 1/8	17	1 5/8

STEEL PIPE SUPPORT SPACING IN FEET

Nominal Pipe Size (in.)	Support Spacing (ft.)	Nominal Pipe Size (in.)	Support Spacing (ft.)
1	7	4	14
1 1/2	9	6	17
2	10	8	19
3	12	10	22

COPPER PLUMBING PIPE AND TUBING

When installing copper pipe, sweat fittings are measured by their inside diameter (ID) and compression fittings are measured by their outside diameter (OD). Always use a 50/50 solid-core solder along with a high-quality flux when soldering sweat fittings. DO NOT use a rosin-core type solder.

Type	Characteristics and Guidelines
DWV	**Drain, waste, and vent** is recommended for above-ground use only. Not to be used in pressure applications and install only using sweat fittings. Available in hard type from 1¼-inch to 6-inch sizes.
K	Flexible copper tubing with a thicker wall than Type L and M. Required for all underground installations. Uses include plumbing, heating, steam, gas, and oil where thick-walled tubing is required. Can be used with sweat, flared and compression fittings. Available in hard and soft types.
L	Standard copper tubing used for interior, above-ground applications including air conditioning, heating, steam, gas, and oil. Because of its flexibility, be very careful not to crimp the line when bending. Tools are available to make bending safer and easier. Sweat, compression, and flare fittings are available. DO NOT use compression fittings for gas lines. Available in hard and soft types.
M	Generally used with interior heating and pressure-line applications. Wall thickness is less than types K and L. Install with sweat fittings. Available in hard and soft types.

WATER DEMAND AT INDIVIDUAL OUTLETS	
Outlet	**Flow Rate (gpm)**
Ordinary lavatory faucet	2.00
Bath faucet, ½"	5.00
Laundry faucet, ½"	5.00
Self-closing lavatory faucet	2.50
Sink faucet, ⅜ or ½"	4.50
Sink faucet, ¾"	6.00
Shower head, ½"	5.00
Ballcock in water closet flush tank	3.00
¾" flush valve (15 psi)	15.00
1" flush valve (15 psi)	27.00
1" flush valve (25 psi)	35.00
Drinking fountain jet	.75
Dishwashing machine (domestic)	4.00
Laundry machine, 8 or 16 lb.	4.00
Aspirator (laboratory or operating room)	2.50
Hose bib or sillcock, ½"	5.00

AVERAGE WATER FLOW PER FIXTURE	
Fixture	**Flow Rate (gpm)**
Ordinary basin faucet	2.00
Self-closing basin faucet	2.50
Sink faucet, ⅜"	4.50
Sink faucet, ½"	4.50
Bathtub faucet	6.00
Shower	5.00
Laundry tub cock, ½"	5.00
Ballcock for water closet	3.00
Flushometer valve for water closet	15.00 to 35.00
Flushometer valve for urinal	15.00
Drinking fountain	.75
Sillcock (wall hydrant)	5.00

AVERAGE WATER USAGE PER ACTIVITY	
Activity	**Gallons Used**
Brushing teeth	1.00
Washing hands	2.00
Shower	25.00
Tub bath	36.00
Dishwashing	50.00
Automatic dishwasher cycle	16.00
Washing machine cycle	60.00

PRESSURE AND FLOW RATE PER FIXTURE		
Fixture	Pressure (psi)	Flow (gpm)
Ordinary basin faucet	8	2.0
Self-closing basin faucet	12	2.5
Sink faucet, ⅜"	10	4.5
Sink faucet, ½"	8	5.0
Bathtub faucet	8	4.0
Shower	8	2.0
Laundry tub cock, ½"	8	4.0
Ballcock for water closet	8	3.0
Flush valve for water closet	15	25.0
Flush valve for urinal	15	15.0
50' garden hose and sillcock	8	5.0

BRANCH PIPING FOR FIXTURES

Fixture	Minimum NPS (in.)
Bathtub	½
Combination sink and laundry tray	½
Drinking fountain	⅜
Dishwashing machine (domestic)	½
Kitchen sink (domestic)	½
Kitchen sink (commercial)	¾
Lavatory	⅜
Laundry tray (1 to 3 compartments)	½
Shower (single head)	½
Sink (service, slop)	½
Sink (flushing rim)	¾
Urinal (flush tank)	½
Urinal (¾" flush valve)	¾
Urinal (1" flush valve)	1
Water closet (flush tank)	⅜
Water closet (flush valve)	1
Hose bib	½
Wall hydrant or sillcock	½

DRAIN AND TRAP SIZES FOR FIXTURE UNITS

Fixture	Number of Fixture Units	Drain and Trap Size (in.)
Lavatory (wash basin)	1	1¼
Lavatory (inset)	2	1½
Water closet, flush tank	4	3
Water closet, flushomatic	6	3
Bathtub	2	1½
Shower (single stall)	2	2
Kitchen sink	2	1½
Dishwasher	2	1½
Laundry tub	2	1½
Washing machine	2	2
Service sink	3	2
Drinking fountain	1	1¼
Urinal (stall)	2	2
Bidet	2	1½
Floor drain	2	2

SERVICE AND DISTRIBUTION PIPE SIZING (46 TO 60 PSI)							
Size of Water Meter and Street Service NPS (in.)	Size of Water Service and Distribution Pipes NPS (in.)	Maximum Length of Water Pipe in feet					
		40'	60'	80'	100'	150'	200'
		Number of Fixture Units					
3/4	1/2	9	8	7	6	5	4
3/4	3/4	27	23	19	17	14	11
3/4	1	44	40	36	33	28	23
1	1	60	47	41	36	30	25
1	1 1/4	102	87	76	67	52	44

MINIMUM SIZE FOR FIXTURE SUPPLY

Fixture	Minimum NPS (in.)
Lavatory	⅜
Bidet	⅜
Toilet	⅜
Bathtub	½
Shower	½
Kitchen sink	½
Dishwasher	½
Laundry tub	½
Hose bib	½

INDIVIDUAL, BRANCH AND CIRCUIT VENT SIZING FOR HORIZONTAL DRAIN PIPES

Drain Pipe Size (in.)	Drain Pipe Grade per Foot (in.)	Vent Pipe Size (in.)	Maximum Developed Length of Vent Pipe (ft.)
1½	¼	1¼	Unlimited
1½	¼	1½	Unlimited
2	¼	1¼	290
2	¼	1½	Unlimited
3	¼	1½	97
3	¼	2	420
3	¼	3	Unlimited
4	¼	2	98
4	¼	3	Unlimited
4	¼	4	Unlimited

VENT STACK AND STACK VENT SIZING			
Drain Pipe Size (in.)	Fixture-Unit Load on Drain Pipe	Vent Pipe Size (in.)	Maximum Developed Length of Vent Pipe (ft.)
1½	8	1¼	50
1½	8	1½	150
1½	10	1¼	30
1½	10	1½	100
2	12	1½	75
2	12	2	200
2	20	1½	50
2	20	2	150
3	10	1½	42
3	10	2	150
3	10	3	1,040
3	21	1½	32
3	21	2	110
3	21	3	810
3	102	1½	25
3	102	2	86
3	102	3	620
4	43	2	35
4	43	3	250
4	43	4	980
4	540	2	21
4	540	3	150
4	540	4	580

WET-VENT STACK SIZING	
Quantity of Wet-Vented Fixtures	**Required Stack Size (in.)**
1 to 2 Bathtubs or showers	2
3 to 5 Bathtubs or showers	2½
6 to 9 Bathtubs or showers	3
10 to 16 Bathtubs or showers	4

WET STACK VENT SIZING		
Stack Size (in.)	**Fixture-Unit Load on Stack**	**Maximum Stack Length (ft.)**
2	4	30
3	24	50
4	50	100
6	100	300

VENT SIZE	
Fixture	**Minimum Vent Size (in.)**
Bathtub	1¼
Lavatory	1¼
Domestic sink	1¼
Shower stalls, domestic	1¼
Laundry tray	1¼
Drinking fountain	1¼
Service sink	1¼
Water closet	2

A minimum of one 3 inch vent must be installed.

TRAP-TO-VENT DISTANCES			
Grade on Drain Pipe (in.)	**Drain Size (in.)**	**Trap Size (in.)**	**Maximum Distance between Trap and Vent (ft.)**
¼	1¼	1¼	3½
¼	1½	1¼	5
¼	1½	1½	5
¼	2	1½	8
¼	2	2	9
⅛	3	3	10
⅛	4	4	12

TRAP SIZES	
Appliance or Fixture	**Size (in.)**
Lavatory	1¼
Drinking fountain	1¼
Dental unit or cuspidor	1¼
Bathtub with or without shower	1½
Bidet	1½
Laundry tray	1½
Dishwasher, domestic	1½
Dishwasher, commercial	2
Washing machine	2
Floor drain	2, 3 or 4
Shower stall, domestic	2
Sinks:	
Combination, sink and tray (with disposal unit)	1½
Combination, sink and tray (with one trap)	1½
Domestic, with or without disposal unit	1½
Commercial, flat rim, bar or counter	1½
Circular or multiple wash	1½
Soda fountain	1½
Surgical	1½
Laboratory	1½
Pot or scullery	2
Service sink	2 or 3
Flushrim or bedpan washer	3
Urinals:	
Trough (per 6 ft. section)	1½
Wall-hung	1½ or 2
Stall	2
Pedestal	3
Water closet	3

TYPICAL ROUGH-IN DIMENSIONS

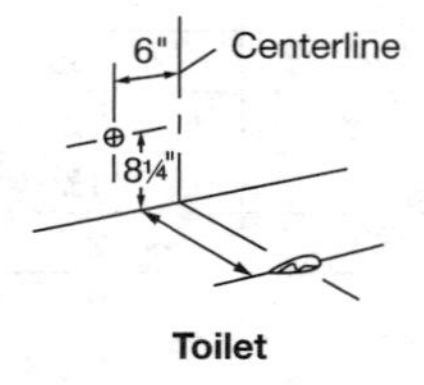

Toilet

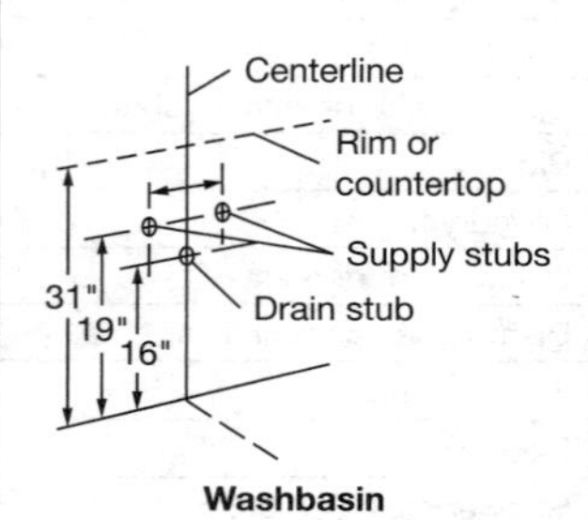

Washbasin

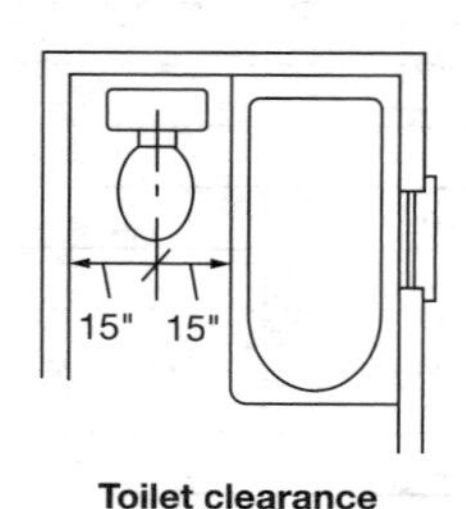

Toilet clearance

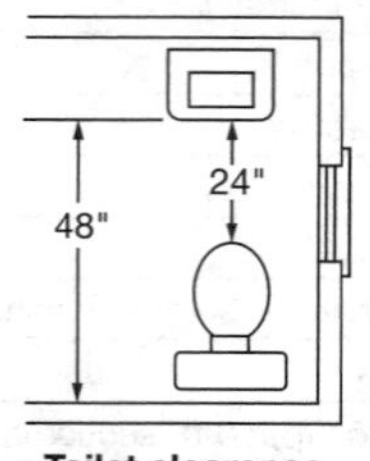

Toilet clearance

TYPICAL ROUGH-IN DIMENSIONS *(cont.)*

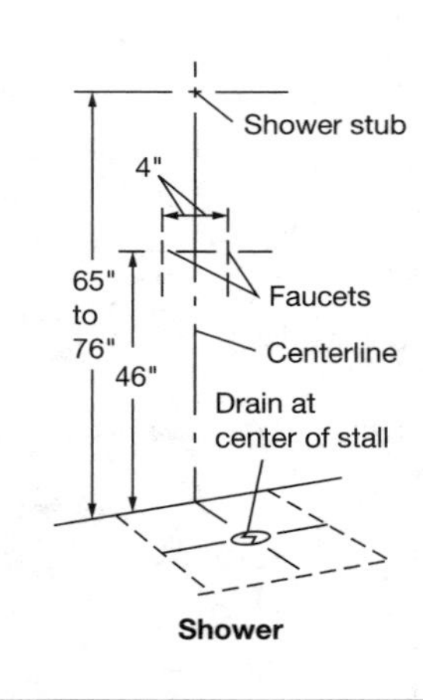

Shower

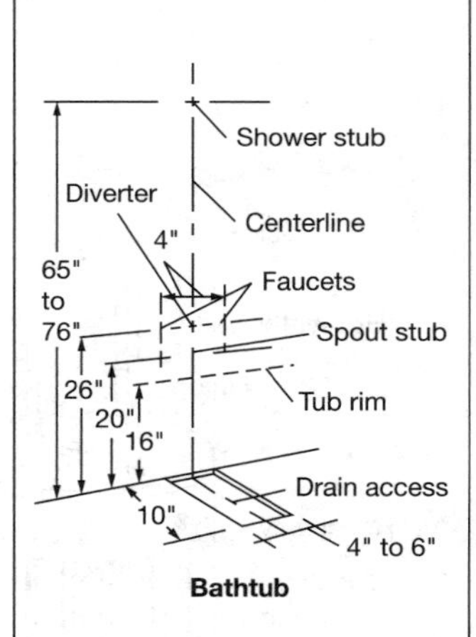

Bathtub

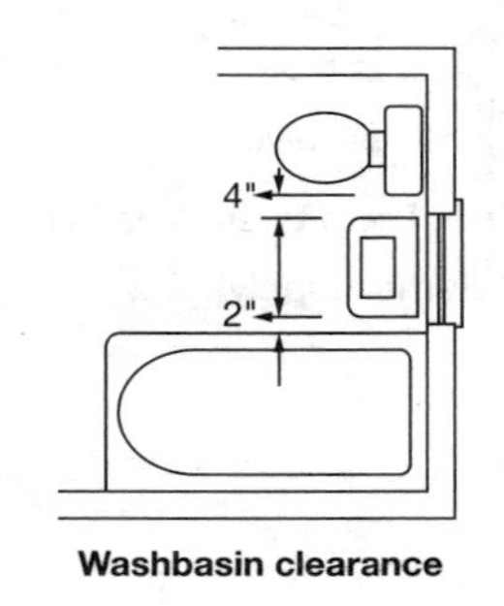

Washbasin clearance

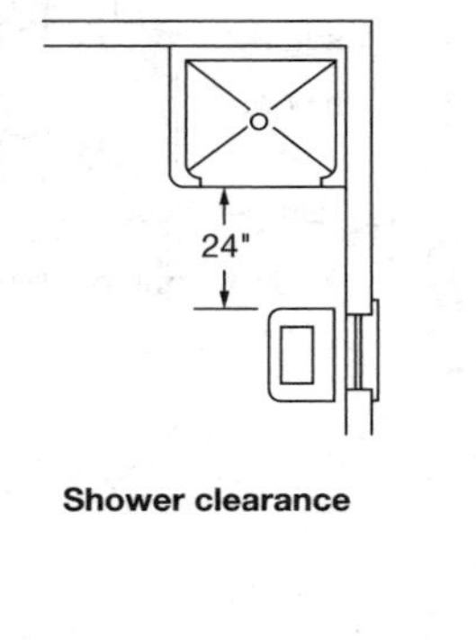

Shower clearance

GAS WATER HEATER

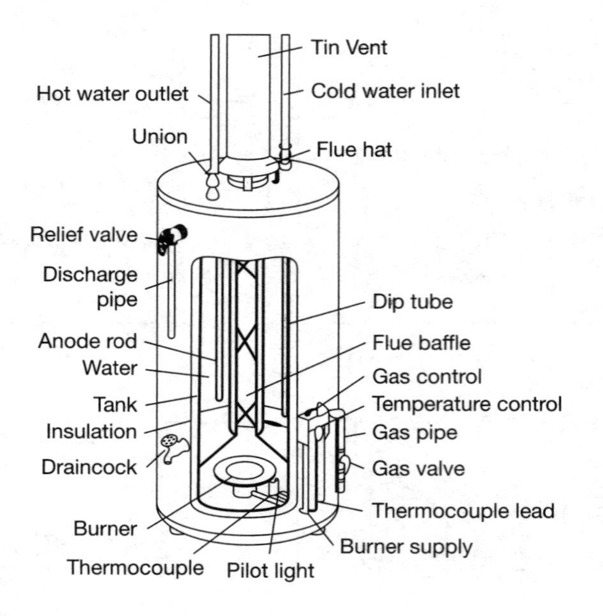

ELECTRIC WATER HEATER

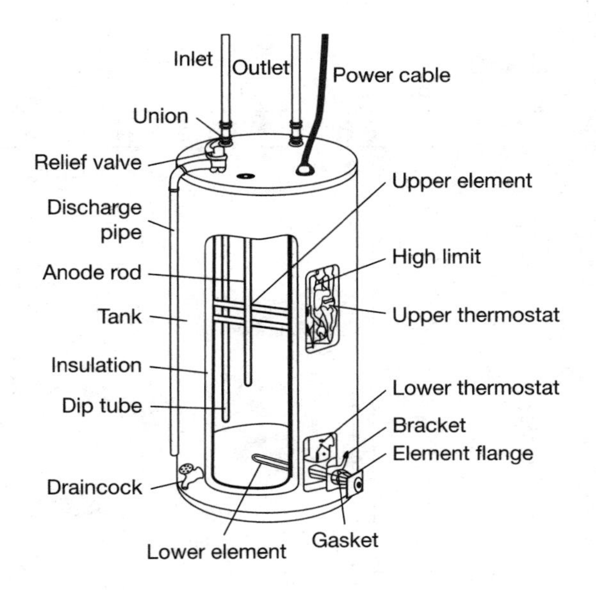

DRIP LEG

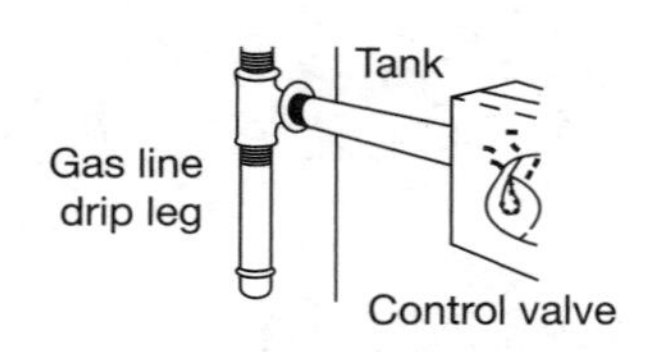

AIR CHAMBER

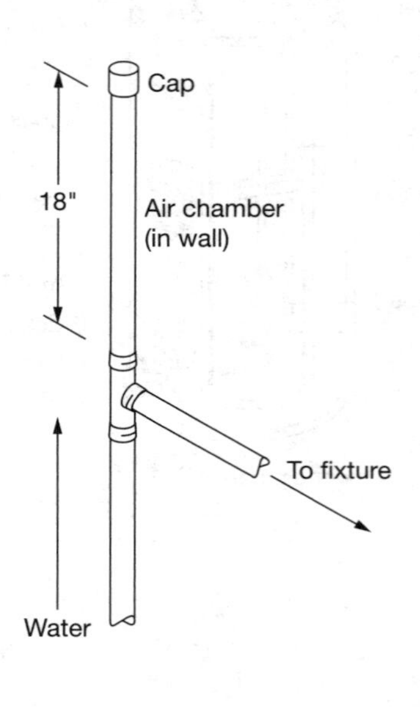

RECOMMENDED MAXIMUM REQUIREMENTS FOR HOT WATER PER DAY IN GALLONS AT 140°F						
Type of Structure	Number of Rooms	Number of Bathrooms				
		1	2	3	4	5
Apartments, Condominiums, Townhouses and Private Homes	1	60	–	–	–	–
	2	70	–	–	–	–
	3	80	–	–	–	–
	4	90	120	–	–	–
	5	100	140	–	–	–
	6	120	160	200	–	–
	7	140	180	220	–	–
	8	160	200	240	250	–
	9	180	220	260	275	–
	10	200	240	280	300	–
	11	–	260	300	340	–
	12	–	280	325	380	450
	13	–	300	350	420	500
	14	–	–	375	460	550
	15	–	–	400	500	600
	16	–	–	–	540	650
	17	–	–	–	580	700
	18	–	–	–	620	750
	19	–	–	–	–	800
	20	–	–	–	–	850

RECOMMENDED HOT WATER TEMPERATURES

Usage	Temperature (F)
Lavatory, hand washing	105°
Lavatory, face washing	115°
Shower and bathtub	110°
Dishwasher and laundry, private	140°
Dishwasher, washing	140°
Dishwasher, sanitizing	180°
Laundry, commercial and institutional	180°
Surgical scrubbing	110°
Commercial buildings, occupant use	110°

MINIMUM RECOMMENDED HOT WATER STORAGE TANK CAPACITY – RESIDENTIAL

Heater Fuel Type	Total Number of Bedrooms	Storage Tank Capacity (gal.)
Gas		20
Electric	1	30
Oil		30
Gas		30
Electric	2	40
Oil		30
Gas		30
Electric	3	50
Oil		30
Gas		40
Electric	4	66
Oil		30

SOIL ABSORPTION CAPACITY/LEACHING AREA		
Type of Soil	Required Square Feet of Leaching Area per 100 Gallons	Maximum Absorption Capacity of Leaching Area per Day in Gal. per Sq. Ft.
Coarse Sand or Gravel	20	5.00
Fine Sand	25	4.00
Sandy Loam or Sandy Clay	40	2.50
Clay with Sand or Gravel	90	1.10
Clay with Small Amount of Sand or Gravel	120	.83

SEPTIC TANK CAPACITY TO LEACHING AREA	
Required Square Feet of Leaching Area per 100 Gallons of Septic Tank Capacity	Allowable Maximum Septic Tank Capacity in Gallons
20 to 25	7,500
40	5,000
60	3,500
90	3,000

SEPTIC TANKS, SINGLE-FAMILY RESIDENCES	
Single Family Residence Number of Bedrooms	**Minimum Septic Tank Capacity (gal.)**
1 to 2	750
3	1,000
4	1,200
5 to 6	1,500

Add 150 gallons per additional bedroom. Septic tank sizes include sludge storage capacity and the connection of food waste disposal units.

SEPTIC TANKS, MULTI-FAMILY RESIDENCES	
Number of Dwelling Units One Bedroom per Unit	**Minimum Septic Tank Capacity (gal.)**
2	1,200
3	1,500
4	2,000
5	2,250
6	2,500
7	2,750
8	3,000
9	3,250
10	3,500

Add 150 gallons for each additional bedroom in a unit.
Add 250 gallons for each dwelling unit over 10.
Septic tank sizes include sludge storage capacity and the connection of food waste disposal units.

CAPACITY OF SEPTIC TANK PER TOTAL FIXTURES

Single-Family Residences Number of Bedrooms	Maximum Fixture Units Served	Minimum Septic Tank Capacity (gal.)
1 to 2	15	750
3	20	1,000
4	25	1,200
5 to 6	33	1,500

DRAINFIELD CONVERSION

Number of Bedrooms	Drainfield Required (Square feet)	Conventional Block Drain (Linear Feet)	Corrugated 4" Plastic Tubing (Linear Feet)
2	100	25	40
3	125	32	50
4	150	38	60
5	175	44	70

SEPTIC TANK DIAGRAM

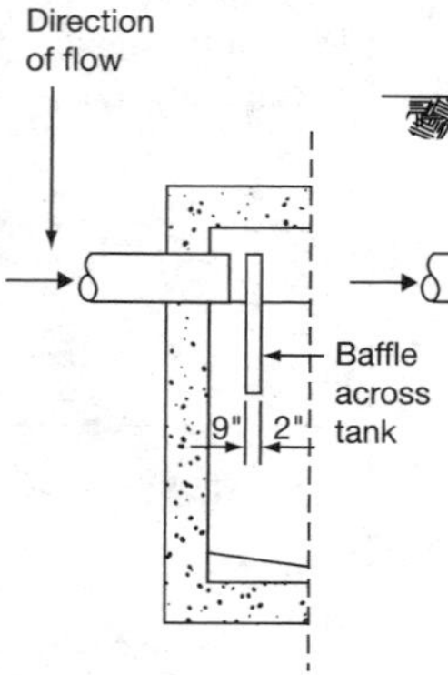

Direction of flow

Baffle across tank

9"

2"

Alternate

with baffle on inlet

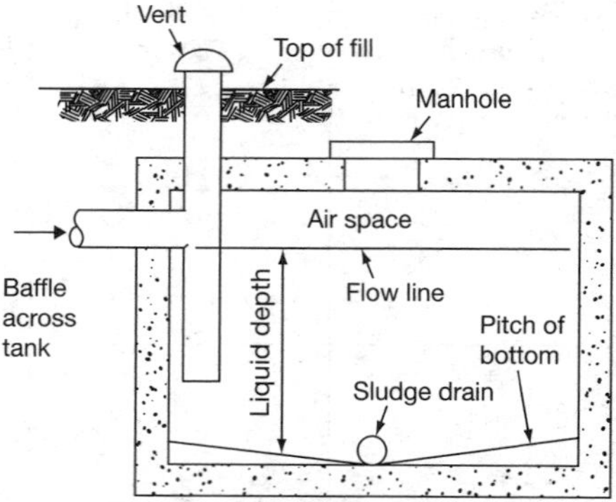

Vent

Top of fill

Manhole

Air space

Flow line

Liquid depth

Pitch of bottom

Sludge drain

Section

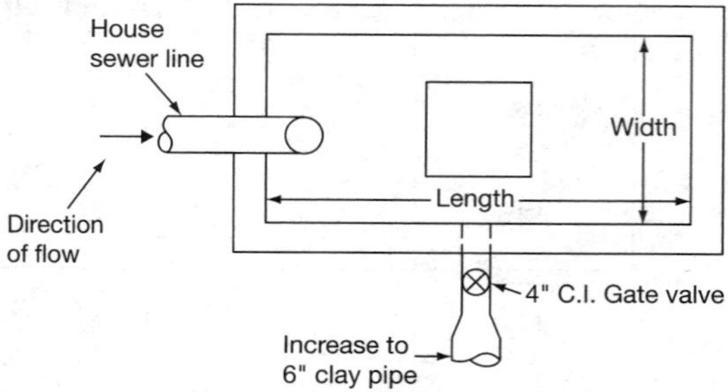

House sewer line

Direction of flow

Width

Length

4" C.I. Gate valve

Increase to 6" clay pipe

Plan

DISTRIBUTION BOX DIAGRAM

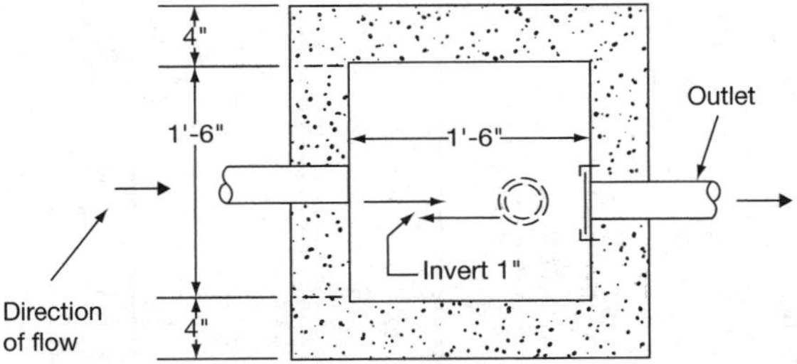

4"
1'-6"
1'-6"
Outlet
Invert 1"
Direction of flow
4"

Section

4"
1'-6"
Invert all outlets 1" lower than inlet
Outlets
Direction of flow
4"
4" 1'-6" 4"

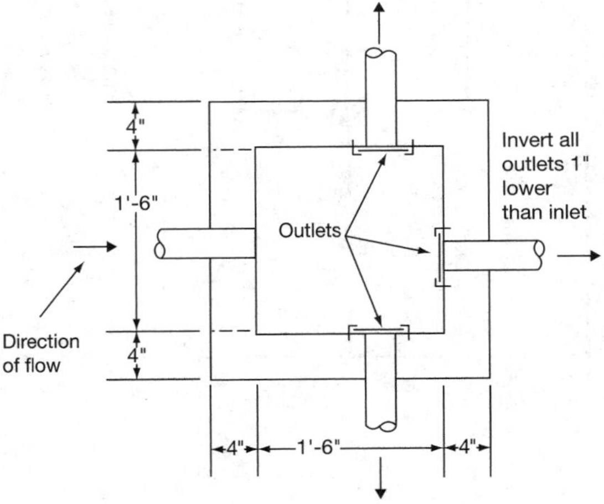

Plan

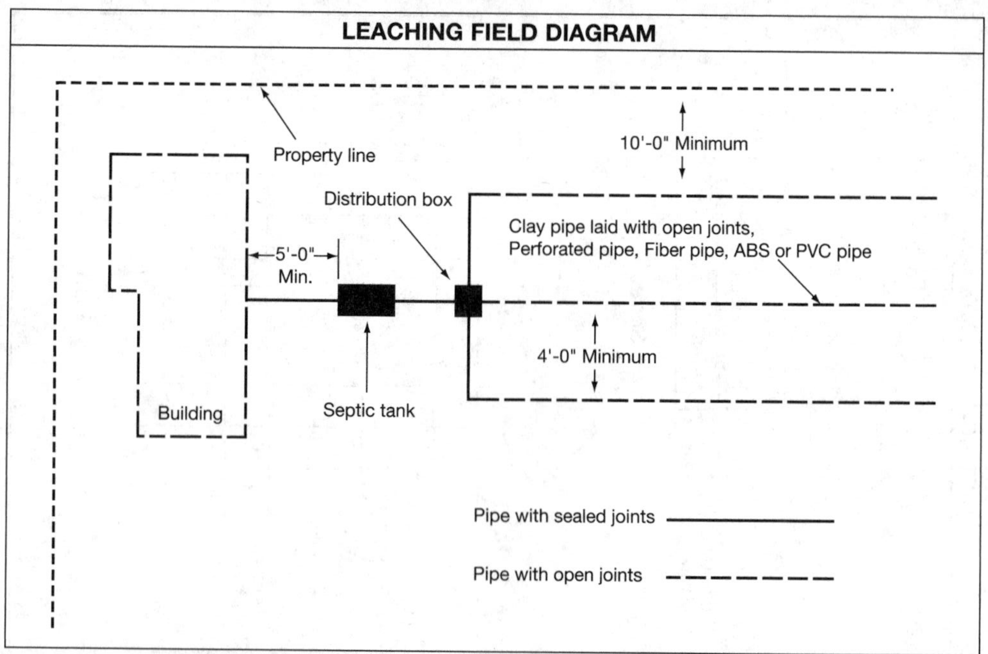
LEACHING FIELD DIAGRAM
Property line
10'-0" Minimum
Distribution box
Clay pipe laid with open joints,
Perforated pipe, Fiber pipe, ABS or PVC pipe
5'-0"
Min.
4'-0" Minimum
Building
Septic tank
Pipe with sealed joints
Pipe with open joints

12 INCH STONE LEACHING CESSPOOL DIAGRAM

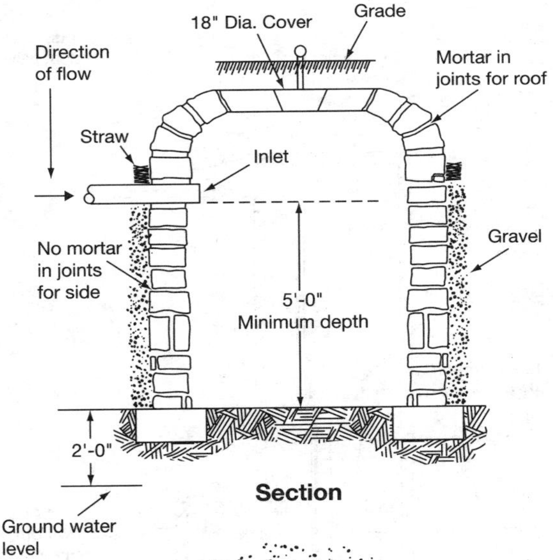

Section

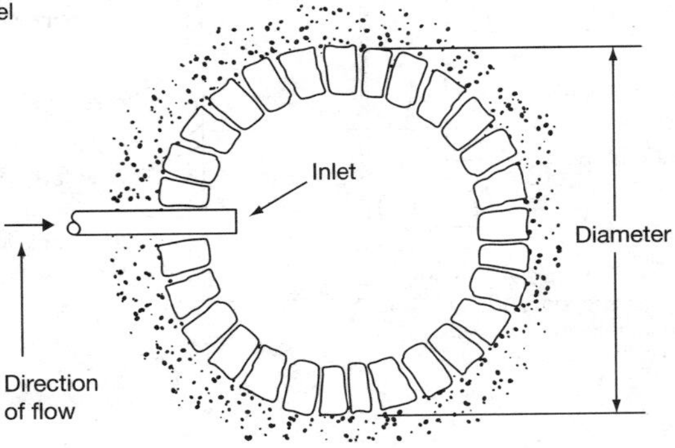

Plan

PUMP TANK DIAGRAM FOR A SAND MOUND

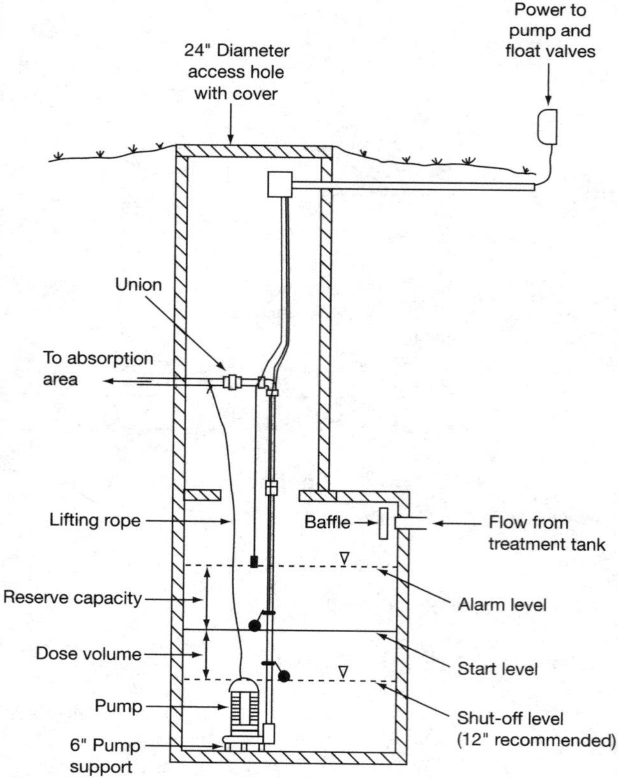

Power to pump and float valves

24" Diameter access hole with cover

Union

To absorption area

Lifting rope

Baffle

Flow from treatment tank

Reserve capacity

Alarm level

Dose volume

Start level

Pump

Shut-off level (12" recommended)

6" Pump support

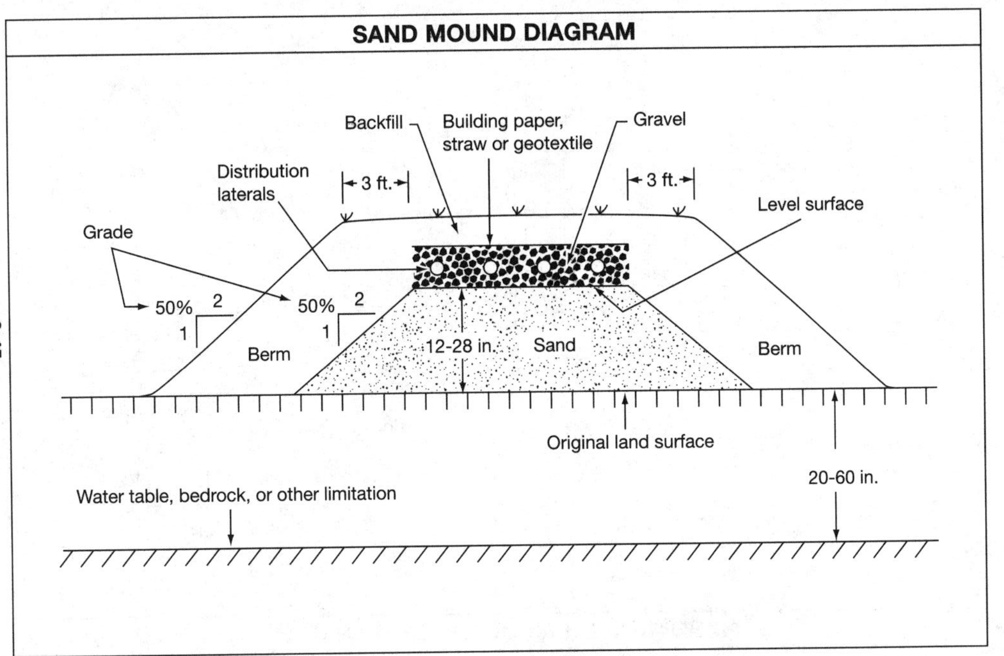
SAND MOUND DIAGRAM
Backfill
Building paper, straw or geotextile
Gravel
Distribution laterals
3 ft.
3 ft.
Level surface
Grade
50%
2
1
50%
2
1
Berm
12-28 in.
Sand
Berm
Original land surface
20-60 in.
Water table, bedrock, or other limitation

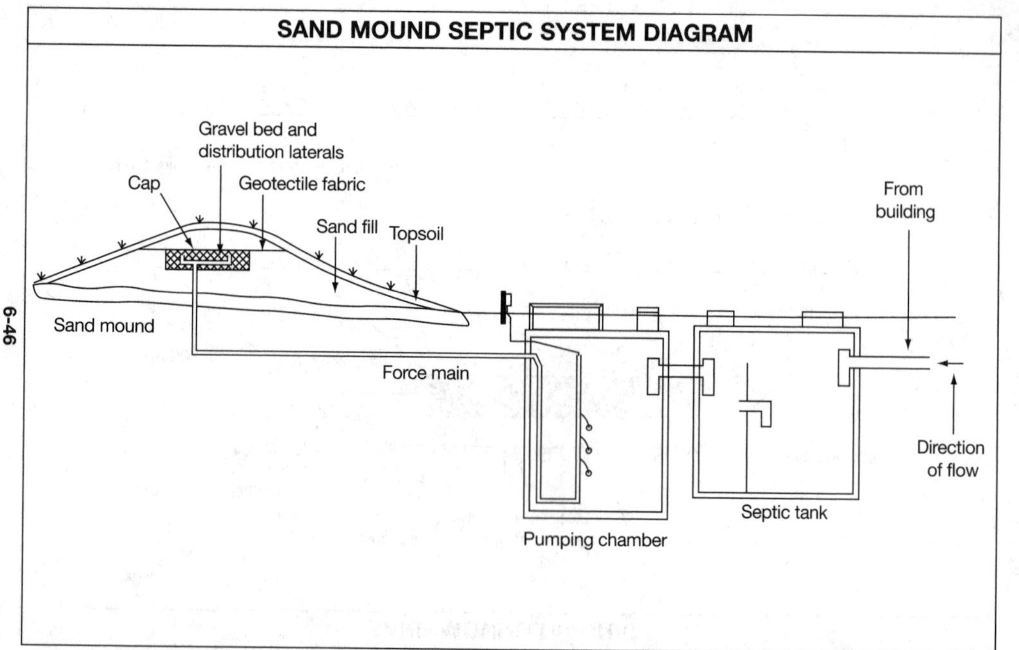
SAND MOUND SEPTIC SYSTEM DIAGRAM
Gravel bed and
distribution laterals
Cap
Geotectile fabric
Sand fill
Topsoil
Sand mound
Force main
From
building
Direction
of flow
Pumping chamber
Septic tank

LOCATION OF SEWAGE DISPOSAL SYSTEM MEASURED IN FEET				
Minimum Horizontal Distance from:	Building Sewer	Septic Tank	Disposal Field	Seepage Pit or Cesspool
Building	2	5	8	8
Private property	Clear	5	5	8
Water supply well	50	50	100	150
Water service line	1	5	5	5
Public water main	10	10	10	10
Distribution box	–	5	5	
Disposal field	–	5	4	5
Seepage pit or cesspool	–	5	5	2
Trees	–	10	–	10
Streams	50	50	50	100
All trenches running parallel and deeper than the footing of a building must be at least 45° from the footing.				

WELL PUMP CHARACTERISTICS

Pump Type	Speed	Suction Lift	Pressure Head	Delivery
Reciprocating Shallow well Low pressure Medium pressure High pressure	Slow 250 to 550 strokes per minute	22 to 25 ft.	40 to 43 psi Up to 100 psi Up to 350 psi	Pulsating Pulsating Pulsating
Deep well	Slow 30 to 50 strokes per minute	Up to 875 ft. (Suction lift below cylinder 22 ft.)	40 psi (normal)	Pulsating
Rotary Pump Shallow well	400 to 1,725 rpm	22 ft.	100 psi (normal)	Slightly pulsating
Ejector Pump Shallow well and limited deep wells	Used with centrifugal turbine or shallow well reciprocating pump	120 ft. (maximum) 80 ft. (normal)	40 psi (normal) 70 psi (maximum)	Continuous, nonpulsating, high capacity, low-pressure head
Centrifugal Shallow well Single stage	High 1,750 and 3,600 rpm	15 ft. (maximum)	40 psi (normal) 70 psi (maximum)	Continuous, non-pulsating, high capacity, low-pressure head
Turbine Type Single impeller	High 1,750 rpm	28 ft. (maximum at sea level)	40 psi (normal) 100 psi (maximum)	Continuous, non-pulsating, high capacity, low-pressure head

SUBMERSIBLE PUMP SELECTION GUIDELINES

Vertical Lift/Elevation

This is the overall vertical distance in feet between the well head and the level where the highest point of water use is located. For example, the vertical distance between the well head and the uppermost shower fixture.

Service Pressure

The service pressure is the range of pressure in the pressure tank during a typical pumping cycle. For example, if a pressure switch with a cut in/cut out pressure rating of 30 psi/50 psi is used, the range of pressure is 40 psi.

Pumping Level

The pumping level is the lowest water level reached in the well during a typical pumping cycle. To measure the pumping level in a well you must first determine the following:

a. Obtain the static (standing) water level in the well. This is the level of water in feet before pumping.

b. Measure the draw-down of the well. This would be the distance in feet that the static water level in the well is lowered by pumping.

Thus, the static water level minus the draw-down will equal the pumping level of the well.

Friction Loss

This is the loss of pressure measured in feet of head due to the resistance to flow inherent in the pipe and fittings.

Total Dynamic Head (TDH)

The TDH is the sum of the vertical lift, service pressure, pumping level, and friction loss and is expressed in feet. TDH and the capacity required determines the pump size.

TOILET FLUSHING ACTIONS

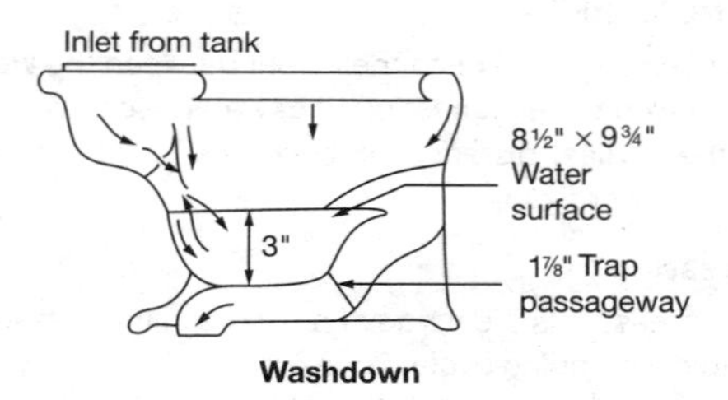

Washdown

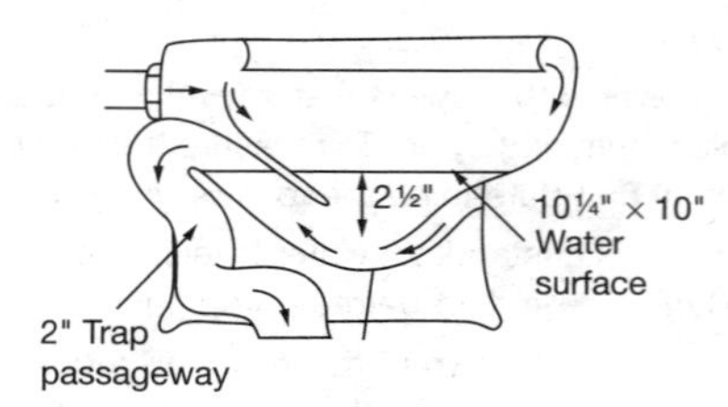

Reverse Trap

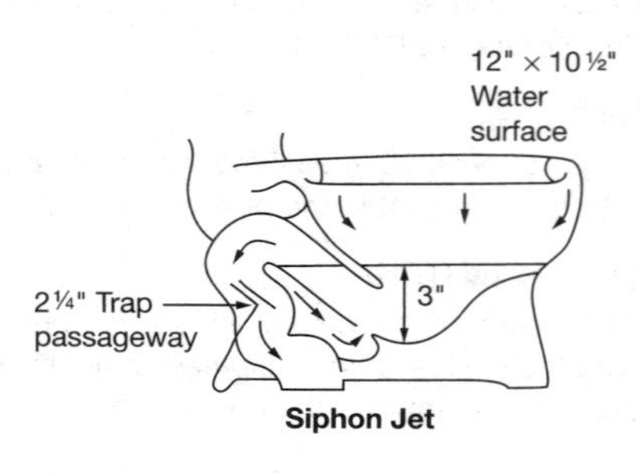

Siphon Jet

TOILET FLUSH MECHANISM

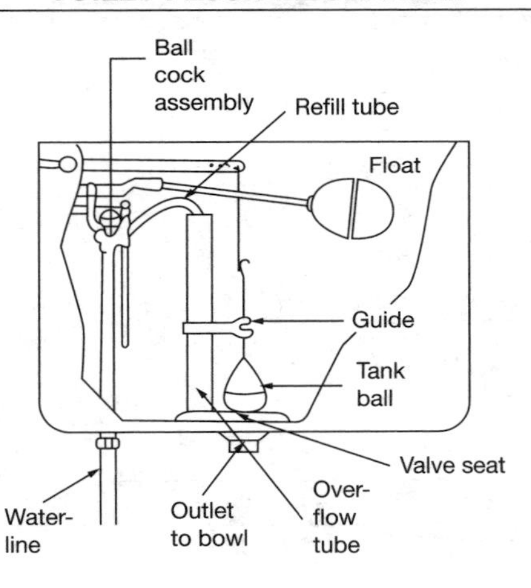

PLUMBING SYMBOLS

Bath Tubs

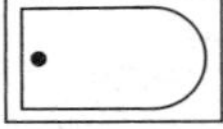

Bathtub, rounded end

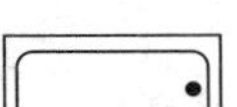

Bathtub, standard

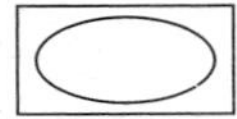

Bathtub, oval

PLUMBING SYMBOLS *(cont.)*	
	Bathtub, whirlpool
	Bidet
	Sitz bath
Grab Bars	
	Grab bar, corner
	Grab bar, straight
Showers	
	Shower head
	Shower stall

PLUMBING SYMBOLS *(cont.)*

Sinks

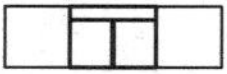

Built in counter double

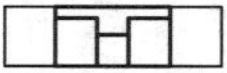

Built in counter triple sink

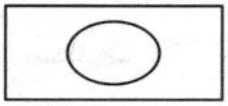

Built in counter

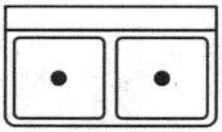

Double kitchen sink

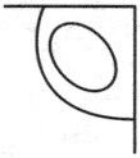

Corner

Floor service sink

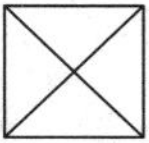

Floor sink

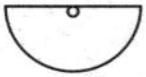

Hand wash sink, half round

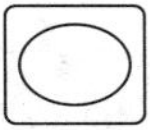

Pedestal

Laundry sink

Service sink

PLUMBING SYMBOLS *(cont.)*

Symbol	Description
	Wall hung
	Wheelchair accessible
Faucets	
	Faucet bath top view
Toilets	
	Toilet, low profile
	Toilet, tank type
	Toilet, floor mounted
	Detention sink and toilet
	Toilet, wall mounted

CHAPTER 7
HVAC

AIR FLOW

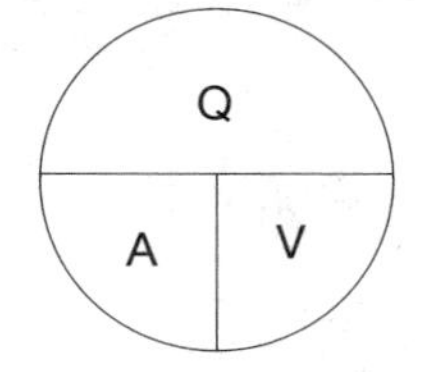

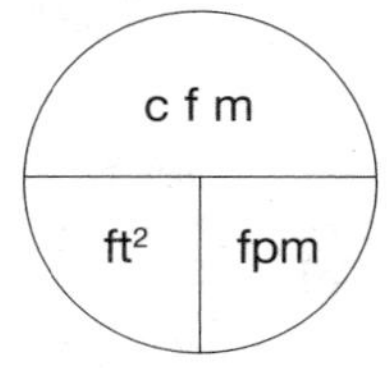

$$cfm = fpm \times area, ft^2$$
$$fpm = cfm/area, ft^2$$
$$area = cfm/fpm$$

AIR CHANGES

$$cfm = space, ft^3 \times \frac{\text{number of changes}}{\text{60 min}}$$

$$\text{number of air changes/hr} = \frac{cfm \times 60 \text{ min}}{space, ft^3}$$

$$\text{cfm per air change} = \frac{space, ft^3}{\text{minutes per change}}$$

FAN LAWS

$$RPM_2 = RPM_1 \times \frac{CFM_2}{CFM_1}$$

$$CFM_2 = CFM_1 \times \frac{RPM_2}{RPM_1}$$

$$SP_2 = SP_1 \times \left(\frac{RPM_2}{RPM_1}\right)^2$$

$$HP_2 = HP_1 \times \left(\frac{RPM_2}{RPM_1}\right)^3$$

PULLEY LAWS

$$\text{rpm of driven} = \frac{\text{diameter of driver} \times \text{rpm}}{\text{diameter of driven}}$$

$$\text{rpm of driver} = \frac{\text{diameter of driven} \times \text{rpm}}{\text{diameter of driver}}$$

$$\text{Diameter of driven} = \frac{\text{diameter of driver} \times \text{rpm}}{\text{rpm of driven}}$$

$$\text{Diameter of the driver} = \frac{\text{diameter of driven} \times \text{rpm}}{\text{rpm of driver}}$$

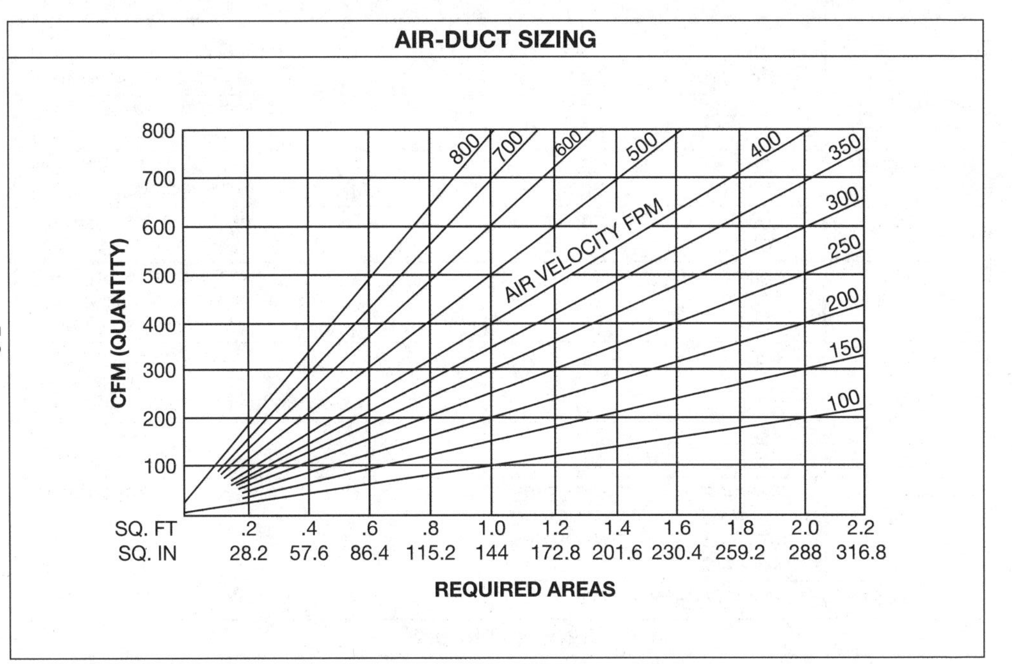
AIR-DUCT SIZING
CFM (QUANTITY)
800
700
600
500
400
300
200
100
SQ. FT
.2
.4
.6
.8
1.0
1.2
1.4
1.6
1.8
2.0
2.2
SQ. IN
28.2
57.6
86.4
115.2
144
172.8
201.6
230.4
259.2
288
316.8
REQUIRED AREAS
AIR VELOCITY FPM
800
700
600
500
400
350
300
250
200
150
100

AIR VELOCITIES

Designation	Recommended Air Velocities FPM		
	Residences, Broadcasting Studios, etc.	Schools, Theaters, Public Buildings	Industrial Applications
Initial air intake	750	800	1,000
Air washers	500	500	500
Extended surface heaters or coolers (face velocity)	450	500	500
Suction connections	750	800	1,000
Through fan outlet			
For 1.5" static pressure	—	2,200	2,400
For 1.25" static pressure	—	2,000	2,200
For 1" static pressure	1,700	1,800	2,000
For 0.75" static pressure	1,400	1,550	1,800
For 0.5" static pressure	1,200	1,300	1,600
Horizontal ducts	700	900	1,000–2,000
Branch ducts and risers	550	600	1,000–1,600
Supply grilles and openings	300	300 grille	400 opening
Exhaust grilles and openings	350	400 grille	500 opening
Duct outlets at high elevation	—	1,000	—

CONVERSION DATA FOR PRESSURES		
Multiply	**Use by**	**To Obtain**
Psi	16	Oz/in.2
Psi	2.31	Ft. H_2O
Psi	27.73	In. H_2O
Psi	0.0703	Kg/cm_2
Psi	2.036	In. Hg
In. H_2O	0.07342	In. Hg
In. H_2O	0.5770	Oz/In.2
In. H_2O	0.03606	Psi
In. H_2O	5.196	Psf
Ft. H_2O	0.4328	Psi
Ft. H_2O	62.32	Psf

RECOMMENDED GAUGES FOR DUCT SYSTEMS

Rectangular ductwork, ½-inch wg static pressure positive or negative, up to 2,000 fpm, based on proper reinforcements spaced at 10-foot intervals.

Largest Dimension, in.	Galvanized Steel Gauge	Aluminum,* B&S Gauge	Copper,* B&S Gauge
Through 26	26	24	24
27–30	24	22	20
31–36	22	20	18
37–48	20	18	18
49–60	18	16	14
73–84	16	14	12
73–84	16	But 8-ft reinforcement spacing required	
85–96	16	But 8-ft reinforcement spacing required	
Over 96	18	But 5-ft class H spacing	

RECOMMENDED GAUGES FOR DUCT SYSTEMS *(cont.)*

Rectangular ductwork, 1-inch wg static pressure positive or negative, up to 2,500 fpm, based on proper reinforcements spaced at 10-foot intervals.

Largest Dimension, in.	Galvanized Steel Gauge	Aluminum,* B&S Gauge	Copper,* B&S Gauge
Through 14	26	24	24
15–24	24	22	20
25–30	22	20	18
31–36	20	18	18
37–42	18	16	14
43–54	16	14	12
55–60	18	But 8-ft reinforcement spacing required	
61–84	18	But 5-ft reinforcement spacing required	
85–96	16	But 5-ft reinforcement spacing required	
Over 96	18	But 2½-ft class H spacing	

RECOMMENDED GAUGES FOR DUCT SYSTEMS *(cont.)*

Rectangular ductwork, 2-inch wg static pressure positive or negative, up to 2,500 fpm

Largest Dimension, in.	Galvanized Steel Gauge	Reinforcement Spacing Intervals, ft.
Through 18	22	10
19–26	20	10
27–30	18	10
31–36	16	10
37–48	16	8
49–60	18	5
61–72	16	5
73–84	18	4, class J
85–96	16	4, class K
Over 96	18	$2\frac{1}{2}$, class H

RECOMMENDED GAUGES FOR DUCT SYSTEMS *(cont.)*		
Rectangular ductwork, 3-inch, wg static pressure positive or negative, up to 4,000 fpm		
Largest Dimension, in.	**Galvanized Steel Gauge**	**Reinforcement Spacing Intervals, ft.**
Through 28	18	10
29–30	16	10
31–36	16	8
37–42	20	5
43–54	18	5
55–60	16	5, class H
61–72	16	4, class I
73–84	18	3, class J
85–96	16	3, class L
Over 96	18	2½, class H

RECOMMENDED GAUGES FOR DUCT SYSTEMS *(cont.)*		
Rectangular ductwork, 4-inch wg static pressure positive, up to 4,000 fpm		
Largest Dimension, in.	**Galvanized Steel Gauge**	**Reinforcement Spacing Intervals, ft.**
Through 12	22	10
13–16	20	10
17–26	18	10
27–30	16	10
31–36	20	5
37–48	18	5
49–54	16	5, class H
55–60	16	5, class I
61–72	18	3, class I
73–84	16	3, class K
85–96	16	2½, class L
Over 96	18	2½, class H with tie rod

RECOMMENDED GAUGES FOR DUCT SYSTEMS *(cont.)*

Rectangular ductwork, 6-inch wg static pressure positive, velocities determined by designer		
Largest Dimension, in.	**Galvanized Steel Gauge**	**Reinforcement Spacing Intervals, ft.**
Through 14	20	10
15–18	18	10
19–22	16	10
23–24	18	8
25–28	16	8
29–36	18	5
37–42	16	5
43–48	18	4
49–54	16	4
55–60	18	3
61–72	16	3
73–84	16	2½
85–96	18	2, class L
Over 96	18	2, class H with tie rod

RECOMMENDED GAUGES FOR DUCT SYSTEMS *(cont.)*

Rectangular ductwork, 10-inch wg static pressure positive, velocities determined by designer

Largest Dimension, in.	Galvanized Steel Gauge	Reinforcement Spacing Intervals, ft.
Through 14	18	8
15–20	16	8
21–28	18	5
29–36	16	5
37–42	16	4
43–48	18	3, class H
49–54	16	3, class I
55–60	16	3, class J
61–72	16	2½, class K
73–84	16	2
Over 85	16	2, class H with tie rod

RECOMMENDED GAUGES FOR DUCT SYSTEMS *(cont.)*

Round Ductwork, Galvanized Steel, Gauge Selection						
Duct Diameter, in.	Maximum 2-in. wg Static Positive		Maximum 10-in. wg Static Positive		Maximum 2-in. wg Static Negative	
	Spiral Seam Gauge, in.	Longitudinal Seam Gauge, in.	Spiral Seam Gauge, in.	Longitudinal Seam Gauge, in.	Spiral Seam Gauge, in.	Longitudinal Gauge, in.
3–8	28	28	26	24	28	24
9–14	28	26	26	24	26	24
15–26	26	24	24	22	24	22
27–36	24	22	22	20	22	20
37–50	22	20	20	20	20	18
51–60	20	18	18	18	18	16
61–84	18	16	18	16	16	14

RECOMMENDED GAUGES FOR DUCT SYSTEMS *(cont.)*

Round Ductwork, Aluminum, Gauge Selection				
Duct Diameter, in.	Maximum 2-in. wg Static Positive		Maximum 10-in. wg Static Positive	
	Spiral Seam Gauge, in.	Longitudinal Seam Gauge, in.	Spiral Seam Gauge, in.	Longitudinal Seam Gauge, in.
3–8	0.025	0.032	0.025	0.040
9–14	0.025	0.032	0.032	0.040
15–26	0.032	0.040	0.040	0.050
27–36	0.040	0.050	0.050	0.063
37–50	0.050	0.063	0.063	0.071
51–60	0.063	0.071	N.A.	0.090
61–84	N.A.	0.090	N.A.	N.A.

CHIMNEY CONNECTOR AND VENT CONNECTOR CLEARANCE FROM COMBUSTIBLE MATERIALS	
Description of Appliance	**Minimum Clearance, Inches**
Residential Appliances	
Single-Wall, Metal Pipe Connectors	
Electric, Gas, and Oil Incinerators	18
Oil and Solid-Fuel Appliances	18
Oil Appliances Listed as Suitable for Use with Type L Venting Systems, but only when connected to chimneys	9
Type L Venting-System Piping Connectors	
Electric, Gas, and Oil Incinerators	9
Oil and Solid-Fuel Appliances	9
Oil Appliances Listed as Suitable for Use with Type L Venting Systems	As listed
Commercial and Industrial Appliances	
Low-Heat Appliances	
Single-Wall, Metal Pipe Connectors	
Gas, Oil, and Solid-Fuel Boilers, Furnaces, and Water Heaters	18
Ranges, Restaurant Type	18
Oil Unit Heaters	18
Other Low-Heat Industrial Appliances	18
Medium-Heat Appliances	
Single-Wall, Metal Pipe Connectors	
All Gas, Oil, and Solid-Fuel Appliances	36

STANDARD CLEARANCE FOR HEAT-PRODUCING APPLIANCES IN RESIDENTIAL INSTALLATIONS

Residential Type Appliances for Installation in Large Rooms		Appliance				
		Above Top of Casing or Appliance	From Top and Sides of Warm-Air Bonnet or Plenum	From Front	From Back	From Sides
Boilers and Water Heaters Steam Boilers–15 psi; Water Boilers–250°F; Water Heaters–200°F; All Water Walled or Jacketed	Automatic Oil or Comb. Gas-Oil	6	–	24	6	6
Furnaces–Central Gravity, Upflow, Downflow, Horizontal And Duct. Warm-Air–250°F Max.	Automatic Oil or Comb. Gas-Oil	6 or as listed	6 or as listed	24	6	6
Furnaces–Floor For Mounting in Combustible Floors	Automatic-Oil or Comb. Gas-Oil	36	–	12	12	12

CLEARANCE (IN INCHES) WITH SPECIFIED FORMS OF PROTECTION

Input Heat Units	Eff. %	Usable Btuh	gph 100°F Rise	Tank Size Gal.	Available Hot-Water Storage Plus Recovery 100°F Rise				Continuous Draw, gph
					15 Min.	30 Min.	45 Min.	60 Min.	
Electricity, kWh									
1.5	92.5	4,750	5.7	20	21.4	22.8	24.3	25.7	5.7
2.5	92.5	7,900	9.5	20	32.4	34.8	37.1	39.5	9.5
4.5	92.5	14,200	17.1	50	54.3	58.6	62.9	67.1	17.1
6.0	92.5	19,000	22.8	66	71.6	77.2	82.8	88.8	22.8
7.0	92.5	22,100	26.5	80	86.6	93.2	99.8	106.5	26.5
Gas, btuh									
3,4000	75	25,500	30.6	30	37.7	45.3	53.0	55.6	25.6
42,000	75	31,600	38.0	30	39.5	49.0	58.8	61.7	31.7
50,000	75	37,400	45.0	40	51.3	62.6	73.9	77.6	37.6
60,000	75	45,000	54.0	50	63.5	77.0	90.5	95.0	45.0
Oil, gph									
0.50	75	52,500	63.0	30	45.8	61.6	77.4	82.5	52.5
0.75	75	78,700	94.6	30	53.6	77.2	100.8	109.0	79.0
0.85	75	89,100	107.0	30	57.7	83.4	110.1	119.1	89.0
1.00	75	105,000	126.0	50	81.5	113.0	144.5	155.0	105.0
1.20	75	126,000	151.5	50	87.9	125.8	163.7	176.0	126.0
1.35	75	145,000	174.0	50	93.5	137.0	180.5	195.0	145.0
1.50	75	157,000	188.5	85	132.1	179.2	226.3	242.0	157.0
1.65	75	174,000	204.5	85	136.1	187.2	238.4	259.0	174.0

PERFORMANCE OF STORAGE WATER HEATERS

Type of Protection	Required Clearance with No Protection											
	36 Inches			18 Inches			12 Inches		9 Inches	6 Inches		
	Above	Sides & Rear	Chimney or Vent Connector	Above	Sides & Rear	Chimney or Vent Connector	Above	Sides & Rear	Chimney or Vent Connector	Above	Sides & Rear	Chimney or Vent Connector
¼ in. asbestos millboard spaced out 1 in.	30	18	30	15	9	12	9	6	6	3	2	3
28-gauge sheet metal on ¼ in. asbestos millboard	24	18	24	12	9	12	9	6	4	3	2	2
28-gauge sheet metal spaced out 1 in.	18	12	18	9	6	9	6	4	4	2	2	2
28-gauge sheet metal on ¼ in. asbestos millboard spaced out 1 in.	18	12	18	9	6	9	6	4	4	2	2	2
1¼ in. asbestos cement covering on heating appliance	18	12	36	9	6	18	6	4	9	2	1	6
¼ in. asbestos millboard on 1 in. mineral fiber batts reinforced with wire mesh or equivalent	18	12	18	6	6	6	4	4	4	2	2	2
22-gauge sheet metal on 1 in. mineral fiber batts reinforced with wire or equivalent	18	12	12	4	3	3	2	2	2	2	2	2
¼ in. asbestos millboard	36	36	36	18	18	18	12	12	9	4	4	4
¼ in. cellular asbestos	36	36	36	18	18	18	12	12	9	3	3	3

INFILTRATION RATE THROUGH VARIOUS TYPES OF WINDOWS

Cubic feet per foot of crack per hour

Window		Wind Velocity, Miles per Hour					
		5	10	15	20	25	30
Double-Hung Wood Sash Windows (Unlocked)	Around frame in masonry wall—not caulked	3	8	14	20	27	35
	Around frame in masonry wall—caulked	1	2	3	4	5	6
	Around frame in wood—frame construction	2	6	11	17	23	30
Double-Hung Metal Windows	Non-weather-stripped, locked	20	45	70	96	125	154
	Non-weather-stripped, unlocked	20	47	74	104	137	170
	Weather-stripped, unlocked	6	19	32	46	60	76
Rolled Section Steel Sash Windows	Industrial pivoted, $^1/_{16}$-in. crack	52	108	176	244	304	372
	Architectural projected, $^1/_{32}$-in. crack	15	36	62	86	112	139
	Architectural projected, $^3/_{64}$-in. crack	20	52	88	116	152	182
	Residential casement, $^1/_{64}$-in. crack	6	18	33	47	60	74
	Residential casement, $^1/_{32}$-in. crack	14	32	52	76	100	128
	Heavy casement section, projected, $^1/_{64}$-in. crack	3	10	18	26	36	48
	Heavy casement section, projected, $^1/_{32}$-in. crack	8	24	38	54	72	92
Hollow Metal, Vertically Pivoted Window		30	88	145	186	221	242

INFILTRATION THROUGH VARIOUS TYPES OF WALL CONSTRUCTION

Types of Wall	Wind Velocity, Miles per Hour					
	5	10	15	20	25	30
Brick wall 8 in. Plain	2	4	8	12	19	23
Plastered	0.02	0.04	0.07	0.11	0.16	0.24
Plain	1	4	7	12	16	21
13 in. Plastered	0.01	0.01	0.03	0.04	0.07	0.10
Frame Wall, Lath and Plaster	0.03	0.07	0.13	0.18	0.23	0.26

RECOMMENDED LIMIT CONTROL SETTINGS

Hot-water system (gravity) . . . . 180°F
Hot-water system (forced) . . . . 160°F
Warm air (forced) . . . . 200°F
Warm air (gravity) . . . . 300°F
Steam system . . . . "off" 3 lb.—"on" 1 lb.
Vapor system . . . . "off" 4 oz.—"on" 2 oz.

COMMON TERMINAL IDENTIFICATIONS

Letter	Wire Color	Terminal Function
R	Red	Power supply; transformer
W	White	Heating control; heating relay or valve coil
Y	Yellow	Cooling control; cooling contactor coil
G	Green	Fan relay coil
O	Orange	Cooling damper
B	Brown	Heating damper
X	—	Malfunction light
P	—	Heat pump contactor coil
Z	—	Low-voltage fan switch

TWIN-TYPE THERMOSTAT

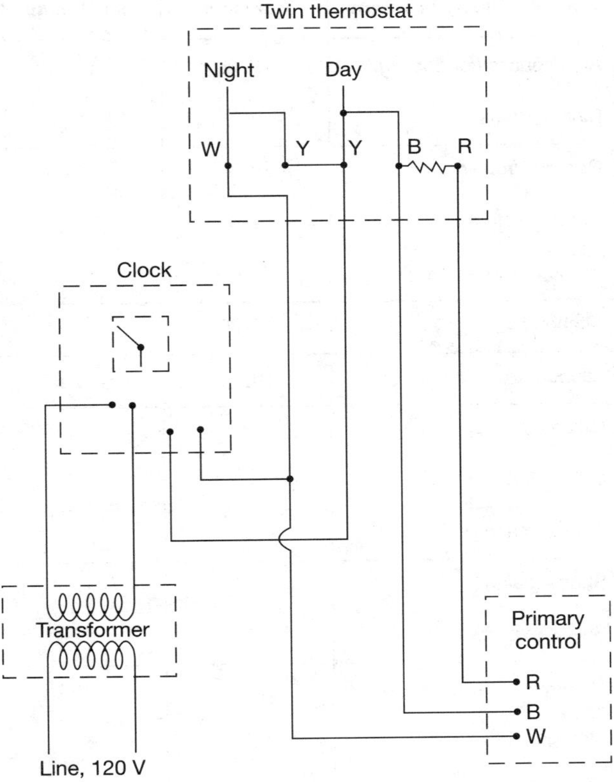

Twin thermostat

Night

Day

W

Y

Y

B

R

Clock

Transformer

Line, 120 V

Primary control

R

B

W

FRESH AIR REQUIREMENTS		
Type of Building or Room	Minimum Air Changes per Hour	Cubic Ft of Air per Minute per Occupant
Attic spaces (for cooling)	12–15	
Dining rooms	5	
Garages (repair)	20–30	
Garages (storage)	4–6	
Homes (night cooling)	9–17	
Kitchens	10–20	
Laundries	10–15	
Offices (public)	3	
Offices (private)	4	
Shops (machine)	5	
Shops (paint)	15–20	
Shops (woodworking)	5	
Theaters		10–15
Storage	2	
Waiting rooms (public)	4	

CHAPTER 8
Electrical

OHM'S LAW/POWER FORMULAS

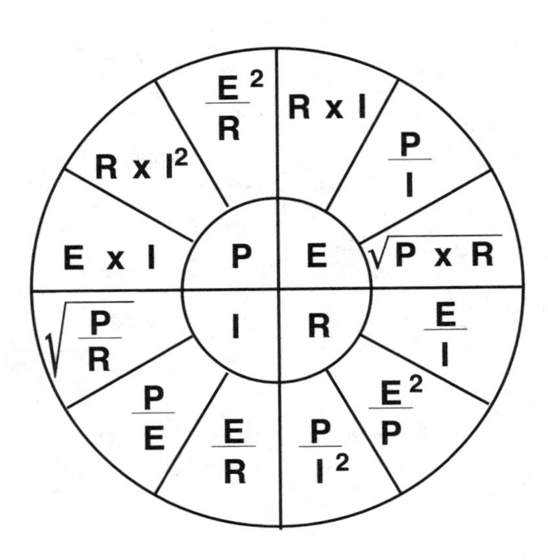

P = Power = Watts

R = Resistance = Ohms

I = Current = Amperes

E = Force = Volts

PULL AND CONNECTION POINTS

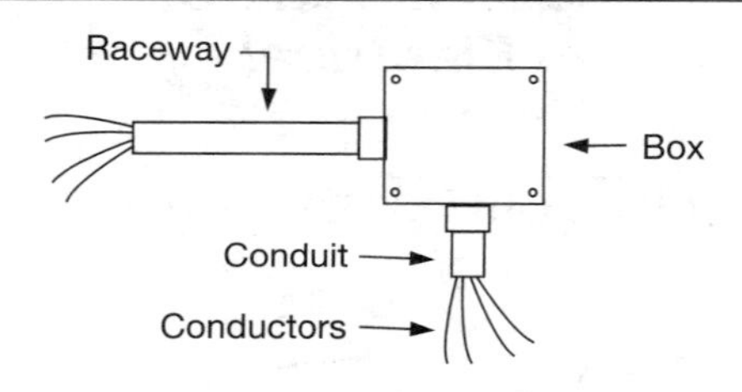

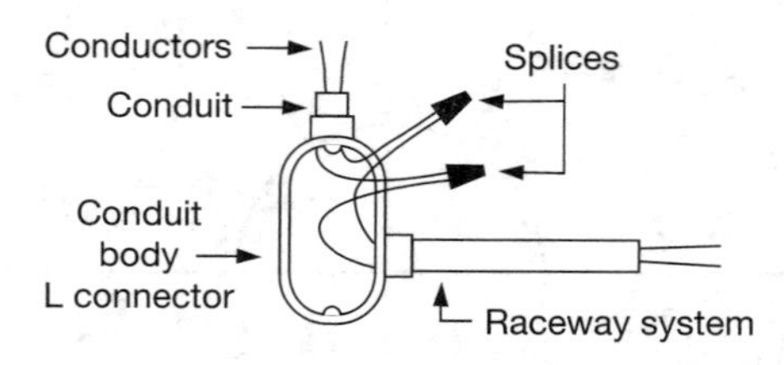

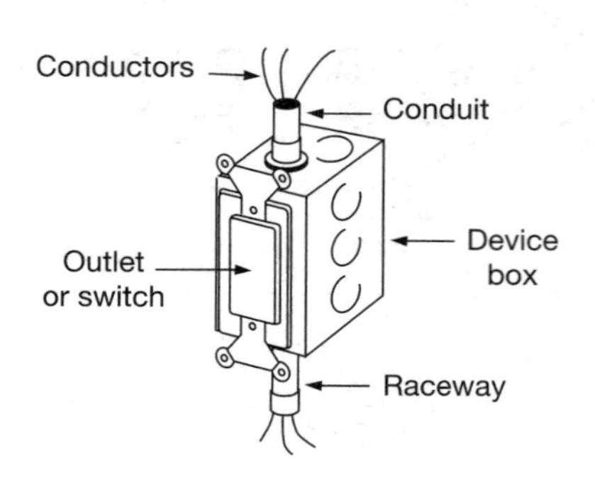

ENCLOSURES IN WET LOCATIONS

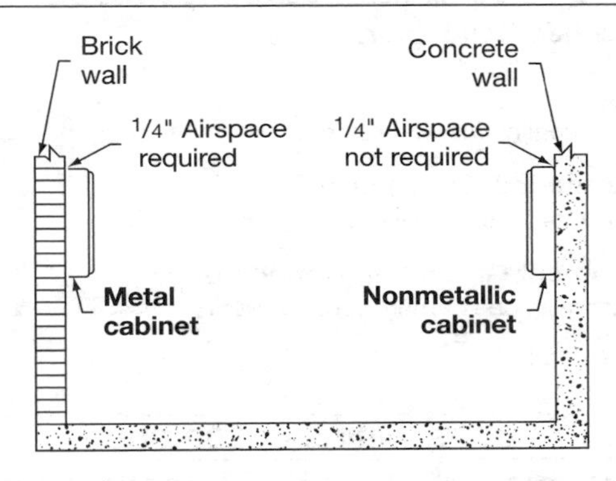

Metal and nonmetallic enclosures mounted in wet locations shall be weatherproof.

GROUNDED CONDUCTOR COLOR CODE

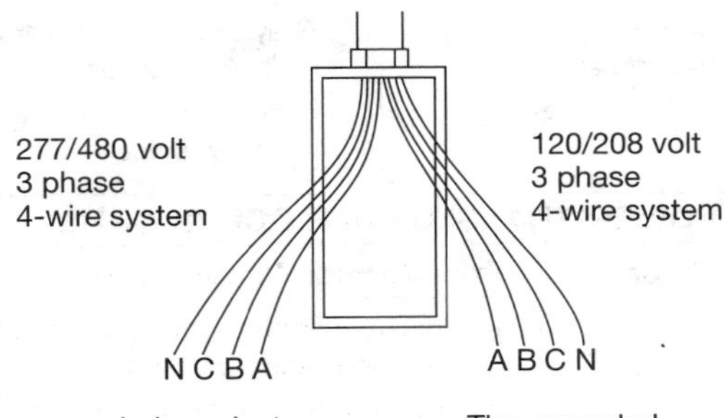

The grounded conductor (same raceway) shall be white with a color stripe, (not green).

The grounded conductor shall be gray or white.

CONDUCTOR COLOR CODE

Grounded Conductor

- White
- Gray
- Three continuous white stripes

Ungrounded Conductor

- Any color other than white, gray, or green

Equipment Grounding Conductor

- Green with one or more yellow stripes
- Bare

POWER WIRING COLOR CODE

120/240 Volt		277/480 Volt	
Black	Phase 1	Brown	Phase 1
Red	Phase 2	Orange	Phase 2
Blue	Phase 3	Yellow	Phase 3
White or with 3 white stripes	Neutral	Gray or with 3 white stripes	Neutral
Green	Ground	Green with yellow stripe	Ground

POWER-TRANSFORMER COLOR CODE

Wire Color	Transformer Circuit Type
Black	If a transformer does not have a tapped primary, both leads are black.
Black	If a transformer does have a tapped primary, the black is the common lead.
Black and Yellow	Tap for a tapped primary.
Black and Red	End for a tapped primary.

NONMETALLIC-SHEATHED CABLE (TYPES NM AND NMC)

- Covered by Article 334 of the National Electrical Code
- Referred to as Romex
- Sizes No. 14 through No. 2 AWG with copper conductors
- Cable will also contain an equipment grounding conductor of proper size.
- Grounding conductor may be bare or covered with green insulation, or green insulation with one or more yellow stripes.
- May be used for both exposed and concealed work in normally dry locations.
- May not be used as service-entrance cable, embedded in poured cement, concrete, or aggregate, in any structure exceeding three floors above grade.
- NMC cable is moisture- and corrosion-resistant, and may also be installed in moist, damp, or corrosive locations.
- NM and NMC cables must be secured within 12 inches of every cabinet, box or fitting and at least every 4½ feet elsewhere.
- NM and NMC cables can be damaged by sharp bending, by being pulled through bored holes, and also by driving the staples holding the cables too forcefully.

METAL-CLAD CABLE (TYPE AC)

- Covered by Article 320 of the National Electrical Code
- Referred to as BX
- AC cables contain a bonding strip.
- Cable may also contain an equipment grounding conductor of proper size.
- Grounding conductor may be bare or covered with green insulation, or green insulation with one or more yellow stripes.
- May be used for both exposed and concealed work in normally dry locations.
- May be used in dry locations, and may be embedded in plaster finish on brick or other masonry, except in damp or wet locations.
- May be run or fished in the air voids of masonry block or tile walls, except where such walls are exposed or subject to excessive moisture or dampness, or are below grade line.
- May not be installed in damp or wet locations, or where exposed to damage.
- AC cables must be secured within 12 inches of every cabinet, box, or fitting and at least every 4½ feet elsewhere.
- AC cables can be damaged by sharp bending.
- At each termination, a fiber bushing must be inserted between the armored sheath and the conductors. The grounding strip must be bent back over the fiber bushing and in intimate contact with the external armor and the cable clamp.

UNDERGROUND FEEDER AND BRANCH-CIRCUIT CABLE (TYPE UF)

- Covered by Article 340 of the NEC®
- Sizes No. 14 to 4/0 AWG with copper conductors
- Cable will also contain an equipment grounding conductor of proper size.
- Grounding conductor may be bare or covered with green insulation, or green insulation with one or more yellow stripes.
- Type UF cable may be buried directly in the earth at a minimum depth of 24 inches. This may be reduced to 18 inches beneath a 2-inch concrete pad, metal raceway, pipe, etc.

SERVICE-ENTRANCE CABLE (TYPES SE AND USE)

- Covered by Article 338 of the NEC®
- Used for service entrances of 100 amps or less.
- May be used for electric ranges.
- Cable will also contain an equipment grounding conductor of proper size.
- Grounding conductor may be bare or covered with green insulation, or green insulation with one or more yellow stripes.
- SC cables must be secured within 12 inches of every cabinet, box, or fitting and at least every 4½ feet elsewhere.

BASIC GROUNDED CONDUCTOR RULES

Circuit breakers or switches shall not disconnect the grounded conductor of a circuit.

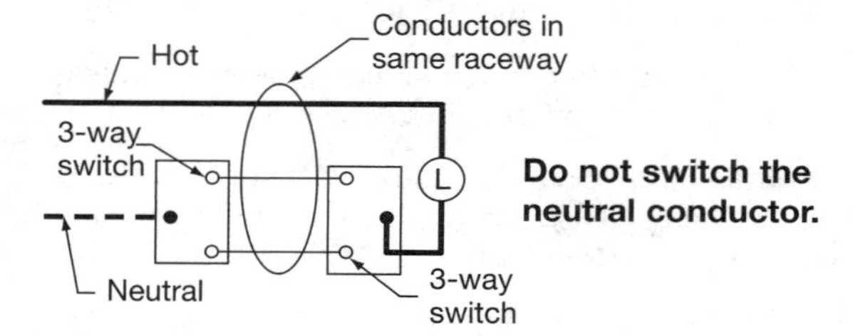

Exception: A circuit breaker or switch may disconnect grounded circuit conductor only if all circuit conductors are disconnected at the same time.

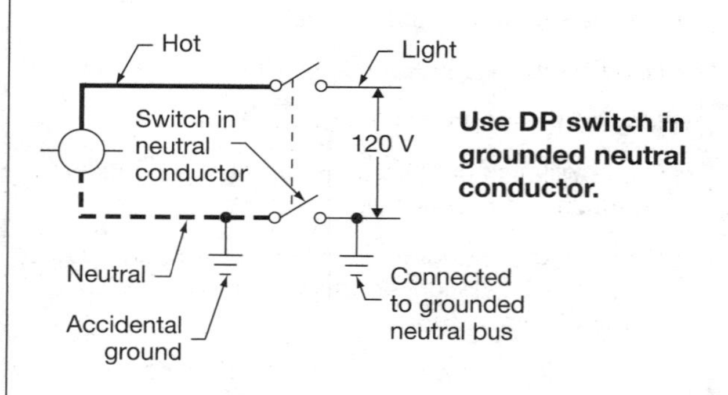

BASIC GROUNDED CONDUCTOR RULES *(cont.)*

Circuit breakers or switches shall not disconnect the grounded conductor of a circuit.

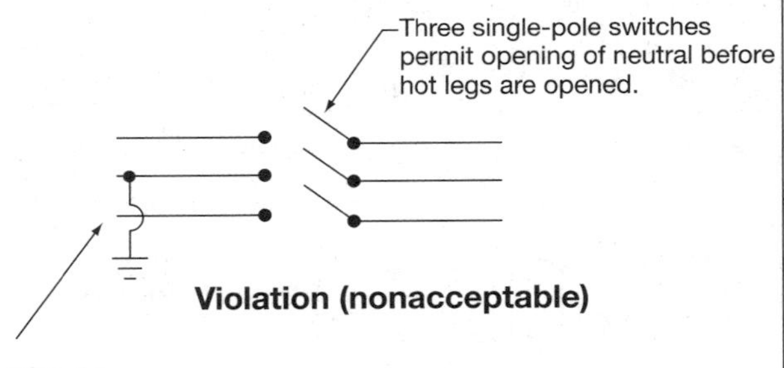

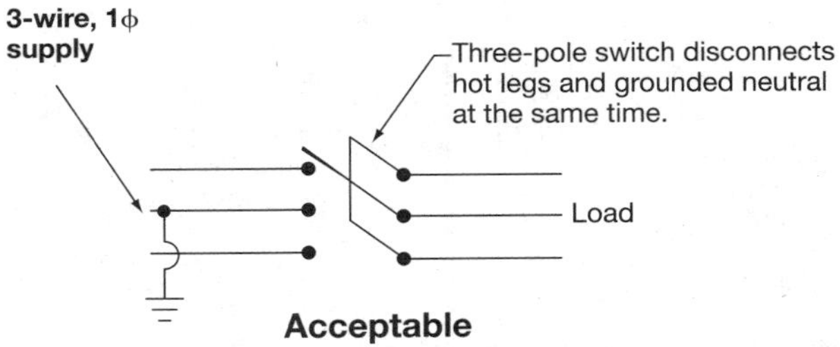

Exception: A circuit breaker or switch may disconnect the grounded circuit conductor if it cannot be disconnected until all other ungrounded conductors have been disconnected.

BASIC GROUNDING CONNECTIONS

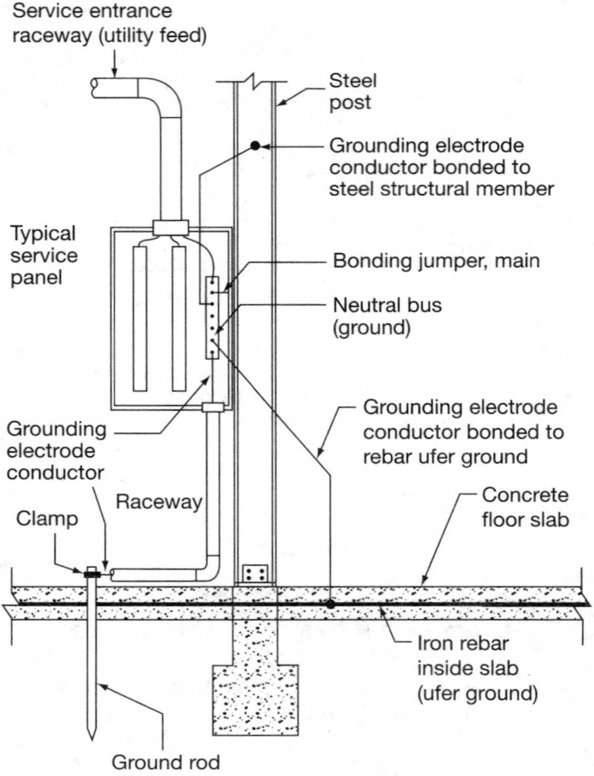

Service entrance raceway (utility feed)

Steel post

Grounding electrode conductor bonded to steel structural member

Typical service panel

Bonding jumper, main

Neutral bus (ground)

Grounding electrode conductor bonded to rebar ufer ground

Grounding electrode conductor

Raceway

Clamp

Concrete floor slab

Iron rebar inside slab (ufer ground)

Ground rod

GROUNDING AND BONDING A TYPICAL SERVICE ENTRANCE

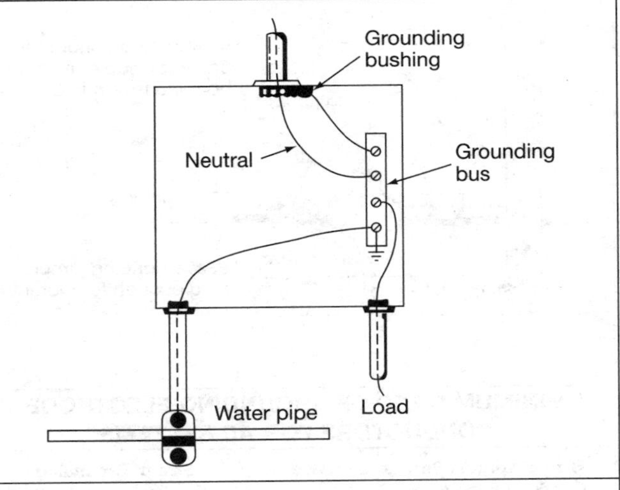

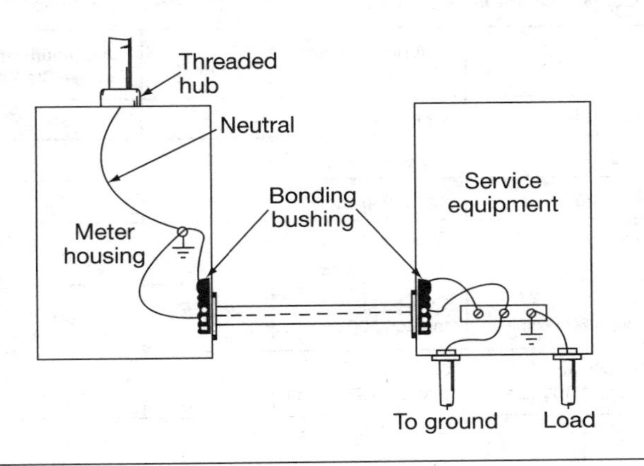

PROPER BONDING OF A WATER METER

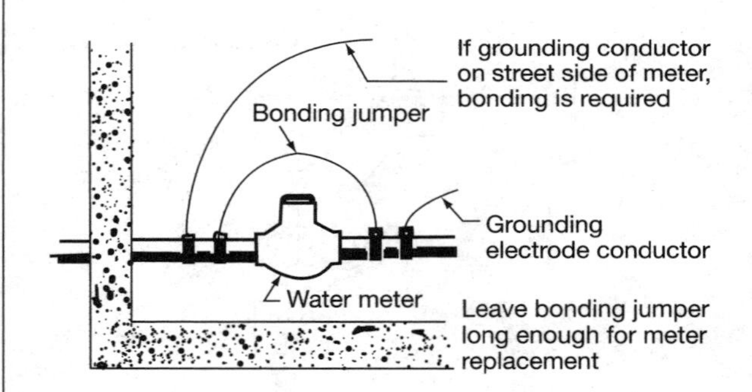

MINIMUM SIZES OF GROUNDING ELECTRODE CONDUCTORS FOR AC SYSTEMS

Size of Largest Service-Entrance Conductor or Parallel Conductors		Size of Grounding Electrode Conductor	
Copper	**Aluminum or Copper-Clad AL**	**Copper**	**Aluminum or Copper-Clad AL**
2 or smaller	0 or smaller	8	6
1 or 2	2/0 or 3/0	6	4
2/0 or 3/0	4/0 or 250 kcmil	4	2
Over 3/0 through 350 kcmil	Over 250 kcmil through 500 kcmil	2	0
Over 350 kcmil through 600 kcmil	Over 500 kcmil through 900 kcmil	0	3/0
Over 600 kcmil through 1,100 kcmil	Over 900 kcmil through 1,750 kcmil	2/0	4/0

MINIMUM SIZE CONDUCTORS FOR GROUNDING RACEWAY AND EQUIPMENT

Setting of Automatic Overcurrent Devices in Circuits Ahead of Equipment, Conduit, etc., Are Not To Exceed the Amperage Ratings Below.	Size (AWG or kcmil)	
	Copper	Aluminum or Copper-Clad Aluminum
15	14 AWG	12 AWG
20	12	10
30	10	8
40	10	8
60	10	8
100	8	6
200	6	4
300	4	2
400	3	1
500	2	1/0
600	1	2/0
800	1/0	3/0
1,000	2/0	4/0
1,200	3/0	250 kcmil
1,600	4/0	350
2,000	250 kcmil	400
2,500	350	600
3,000	400	600
4,000	500	800
5,000	700	1,200
6,000	800	1,200

GROUNDING A TYPICAL RESIDENTIAL WIRING SYSTEM

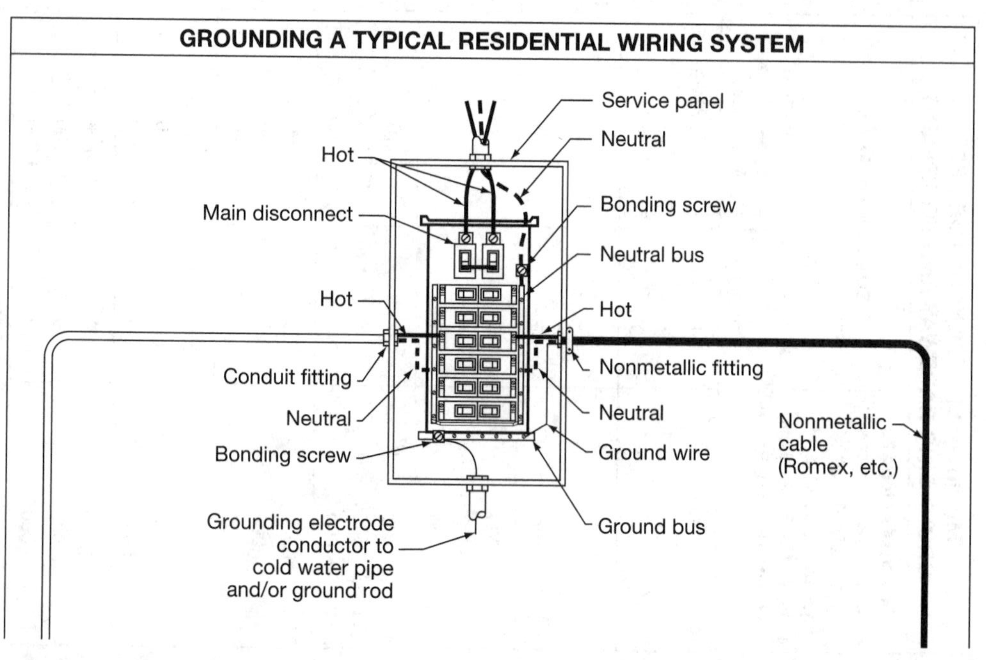

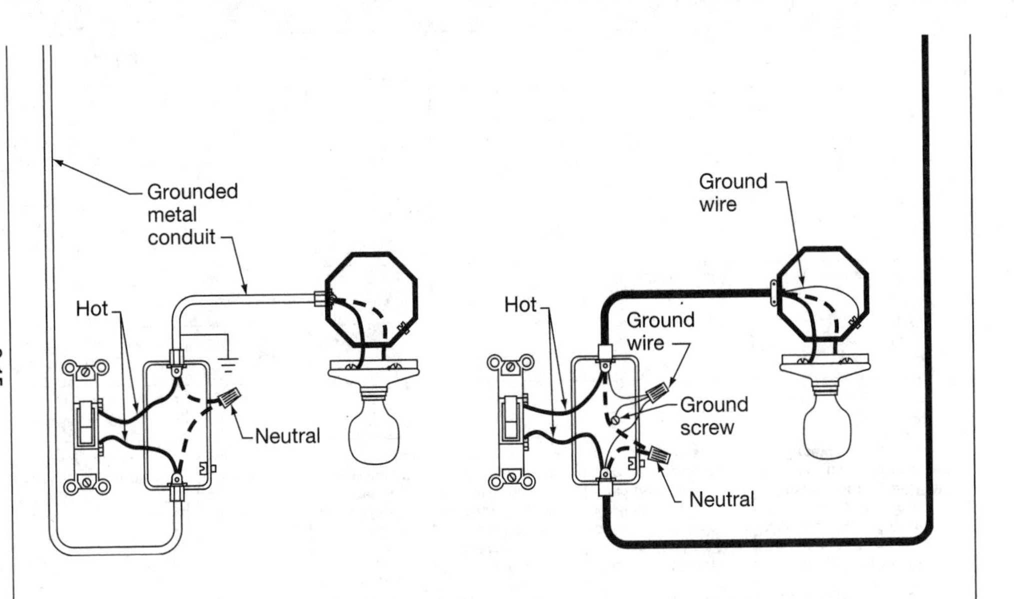
Grounded metal conduit
Hot
Neutral
Ground wire
Hot
Ground wire
Ground screw
Neutral

MINIMUM COVER REQUIREMENTS FOR UNDERGROUND INSTALLATIONS

Location of Circuit or Wiring Method	Type of Circuit or Wiring Method (0 to 600 Volts, Nominal)				
	Direct Burial Cables or Conductors	Rigid Metal Conduit or Intermediate Metal Conduit	Nonmetallic Raceways Listed for Direct Burial without Concrete Encasement or Other Approved Raceways	Residential Branch Circuits Rated 120 V or Less with GFCI Protection and Maximum Over-current Protection of 20 Amperes	Circuits for Control of Irrigation and Landscape Lighting Limited To Not More than 30 V and Installed with Type UF or in Other Identified Cable or Raceway
All locations not specified in these charts	24"	6"	18"	12"	6"
In trench below 2"-thick concrete or equivalent	18"	6"	12"	6"	6"
Under buildings	— (In raceway only)	—	—	— (In raceway only)	— (In raceway only)
Under minimum of 4"-thick concrete exterior slab with no traffic and slab extending no less than 6 inches beyond the installation	18"	4"	4"	6" (Direct burial) 4" (In raceway)	6" (Direct burial) 4" (In raceway)
Under alleys, highways, roads, driveways, streets, and parking lots	24"	24"	24"	24"	24"

MINIMUM COVER REQUIREMENTS FOR UNDERGROUND INSTALLATIONS *(cont.)*

Location of Circuit or Wiring Method	Type of circuit or wiring method (0 to 600 Volts, nominal)				
	Direct Burial Cables or Conductors	**Rigid Metal Conduit or Intermediate Metal Conduit**	**Nonmetallic Raceways Listed for Direct Burial without Concrete Encasement or Other Approved Raceways**	**Residential Branch Circuits Rated 120 V or Less with GFCI Protection and Maximum Over-current Protection of 20 Amperes**	**Circuits for Control of Irrigation and Landscape Lighting Limited To Not More than 30 V and Installed with Type UF or in Other Identified Cable or Raceway**
Under family dwelling driveways. Outdoor parking areas used for dwelling-related purposes only	18"	18"	18"	12"	18"
In or under airport runways including areas where the public is prohibited	18"	18"	18"	18"	18"

Notes: Cover is defined as the shortest distance between a point on the top service of any direct-buried conductor, cable, conduit, or other raceway and the top surface of finished grade, concrete, etc.

Raceways approved for burial only where concrete-encased shall require a 2-inch concrete envelope.

Lesser depths are permitted where cables/conductors rise for terminations, splices, or equipment.

Where solid rock prevents compliance with the cover depths specified in this table, the wiring must be installed in raceways permitted for direct burial. The raceways shall be covered by a minimum of 2 inches of concrete extending down to the solid rock.

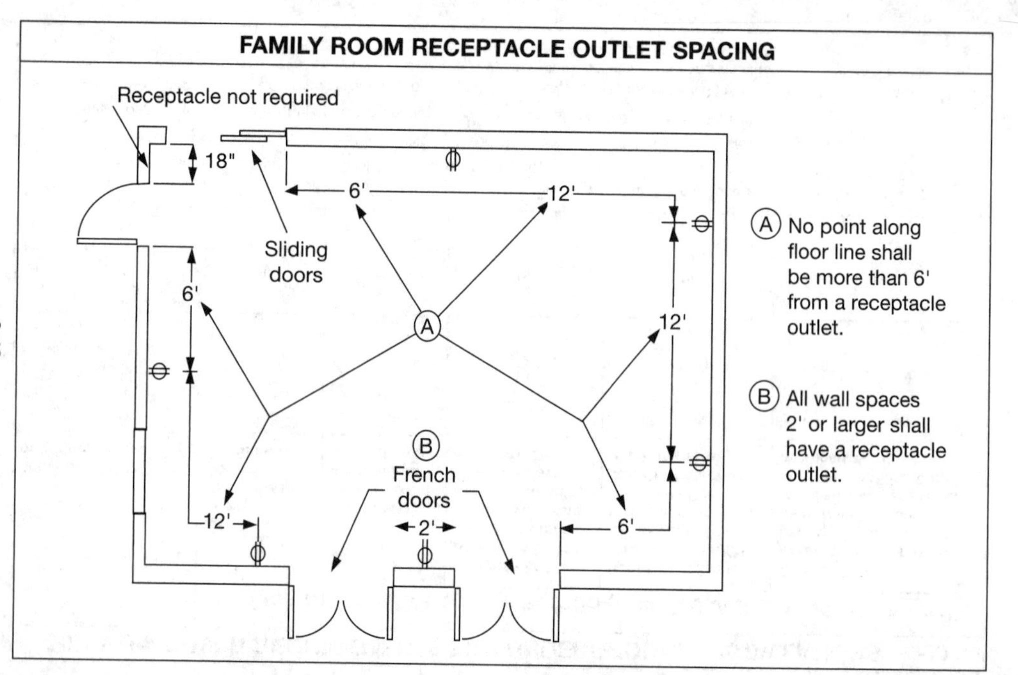
FAMILY ROOM RECEPTACLE OUTLET SPACING
Receptacle not required
18"
Sliding doors
6'
12'
6'
A
12'
12'
6'
B
French doors
2'
A No point along floor line shall be more than 6' from a receptacle outlet.
B All wall spaces 2' or larger shall have a receptacle outlet.

FAMILY ROOM WITH SPLIT-WIRED RECEPTACLES AND SWITCHED CIRCUIT

A split-wired receptacle has the tab between the brass (hot) terminals removed but silver (neutral) terminals remain intact. This provides either a switched circuit or two separate circuits at the same receptacle outlet.

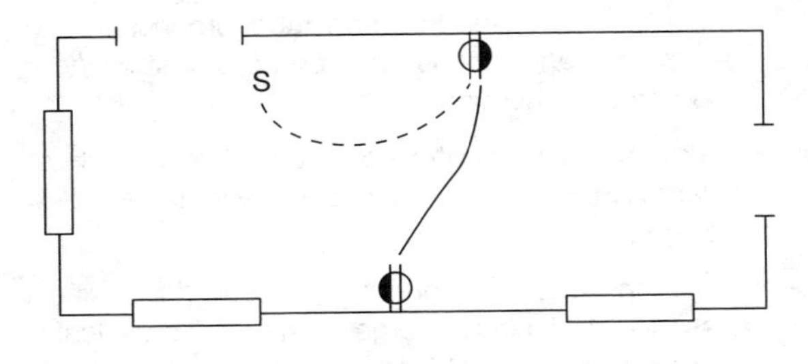

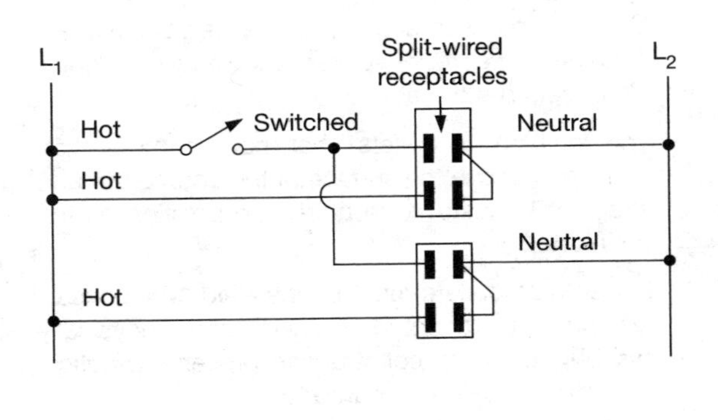

COUNTERTOP RECEPTACLE OUTLET SPACING

The following pertains to receptacle outlets for counterspaces in kitchens, dining rooms, and bar areas of dwelling units.

- A receptacle outlet shall be installed for all countertops that are 12 inches wide or larger.
- There shall be no point along the wall greater than 24 inches from a receptacle outlet. This is due to change in standards for kitchen countertop appliances such as blenders and food processors. A 2-foot cord and plug is required for any appliance used on top of counters.
- For each island countertop that is 24 × 12 inches or larger, there shall be at least one receptacle outlet installed.
- For each peninsula countertop space that measures 24 × 12 inches or larger, there shall be at least one receptacle outlet installed. This measurement is from the connection point between countertops.
- Countertops separated by appliances or sinks are considered separate spaces and require receptacle outlets per the above requirement.
- The location of outlets shall not be more than 18 inches above the surface of the countertop and shall not be installed in a face-up position on the countertop.
- Receptacle outlets must be installed 12 inches or less below the countertop. This is for the physically disabled or when construction prevents practical mounting above the countertop.

COUNTERTOP RECEPTACLE OUTLET SPACING *(cont.)*

Maximum 18"
GFCI
GFCI

24" or greater
Minimum one outlet
ISLAND
12" or greater
GFCI

12" or less below countertop
12" or greater
Minimum one outlet
Countertop extension maximum 6"
24" or greater

Within 4'
GFCI
12" or wider countertops
GFCI

KITCHEN RECEPTACLES

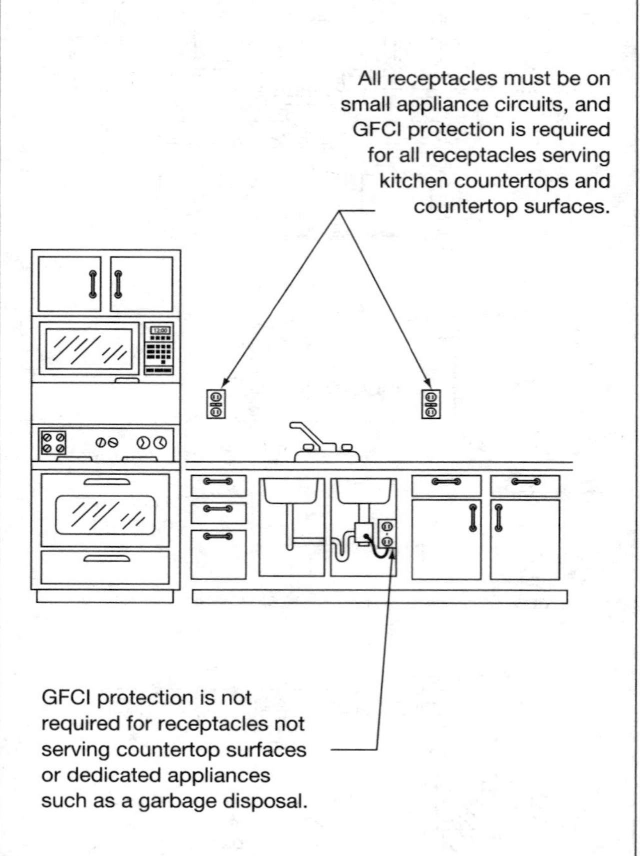

BATHROOM RECEPTACLES

At least one wall receptacle outlet shall be installed within 36" of outside edge of each basin and on a wall adjacent to basin location.

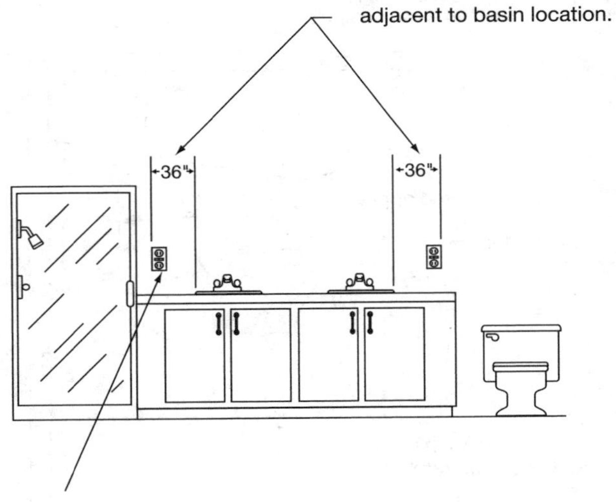

All GFCI-protected receptacles shall be supplied by at least one 20 amp circuit with no other type of outlets on the circuit.

All bathroom receptacles in dwelling units are required to be GFCI-protected.

WET BAR RECEPTACLES

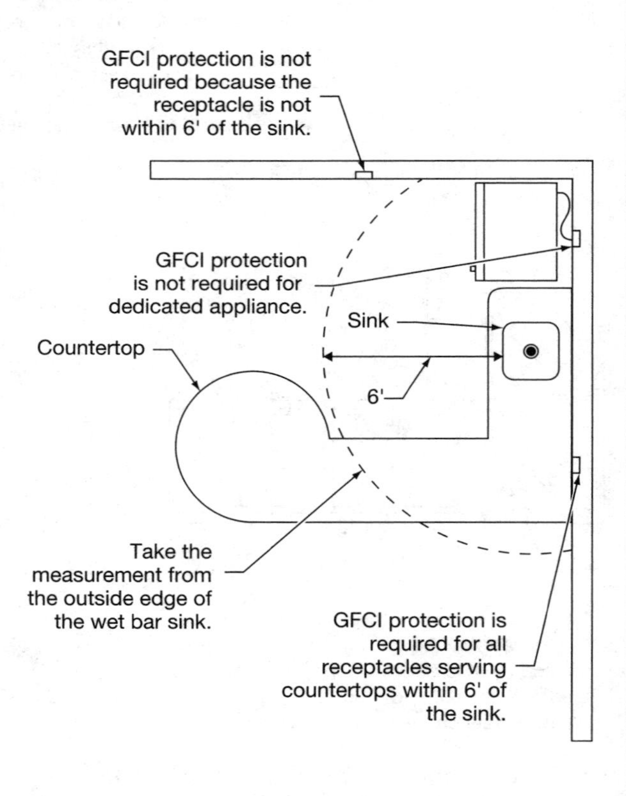

CRAWL SPACE RECEPTACLES

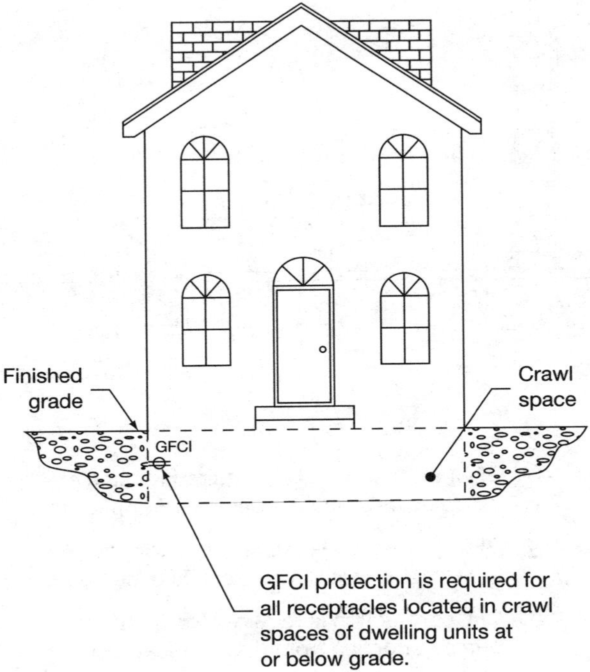

Finished grade

Crawl space

GFCI

GFCI protection is required for all receptacles located in crawl spaces of dwelling units at or below grade.

UNFINISHED BASEMENT RECEPTACLES

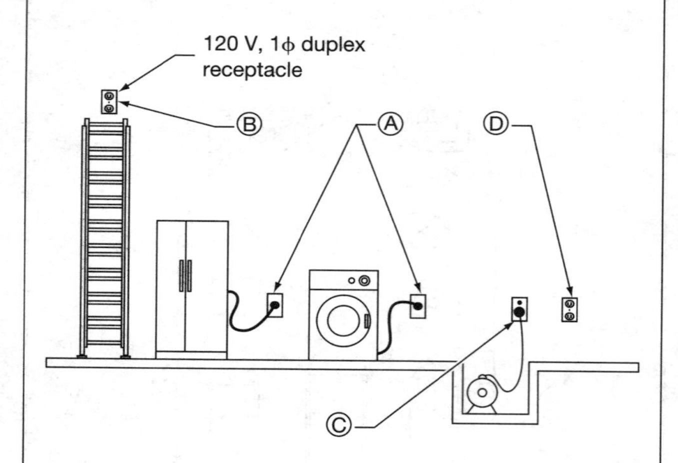

Ⓐ GFCI protection is not required for single receptacles powering dedicated appliances.

Ⓑ GFCI protection is not required for any receptacles that are not readily accessible.

Ⓒ GFCI protection is required for sump pumps supplied from duplex receptacles.

Ⓓ GFCI protection is required for all other duplex receptacles.

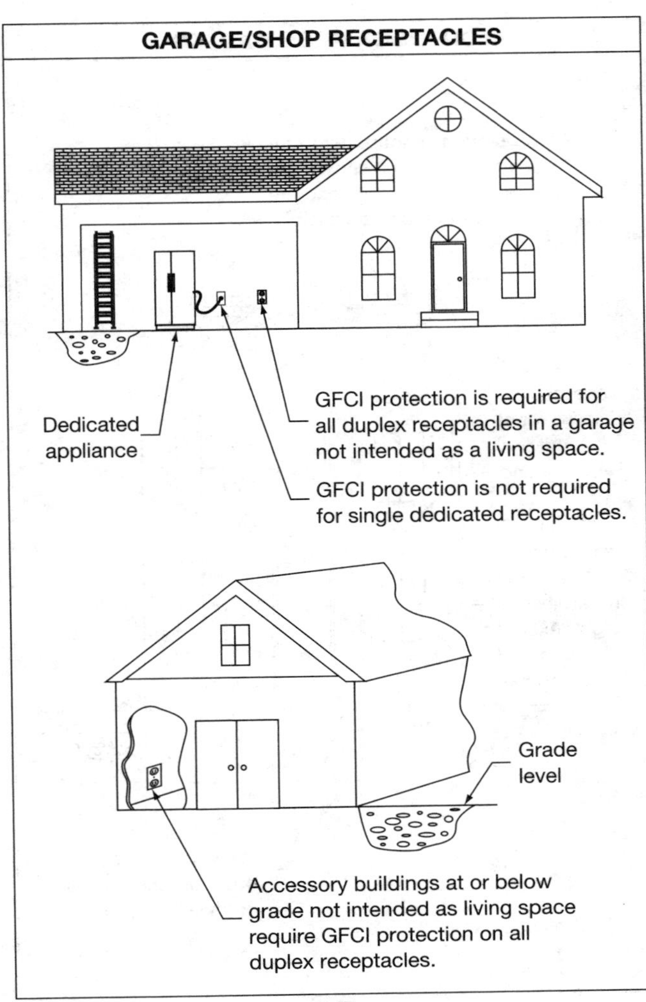
GARAGE/SHOP RECEPTACLES
Dedicated appliance
GFCI protection is required for all duplex receptacles in a garage not intended as a living space.
GFCI protection is not required for single dedicated receptacles.
Grade level
Accessory buildings at or below grade not intended as living space require GFCI protection on all duplex receptacles.

OUTDOOR RECEPTACLES

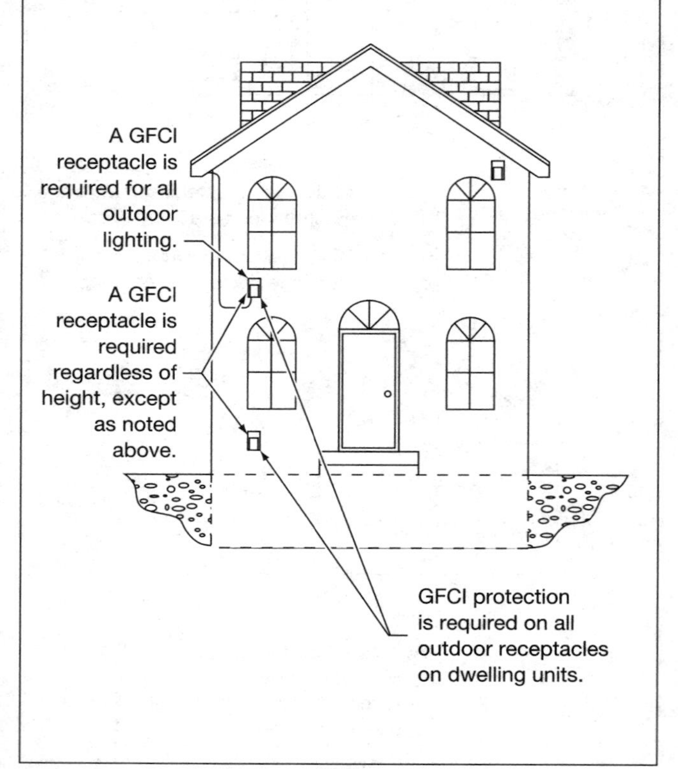

GFCI WIRING DIAGRAMS

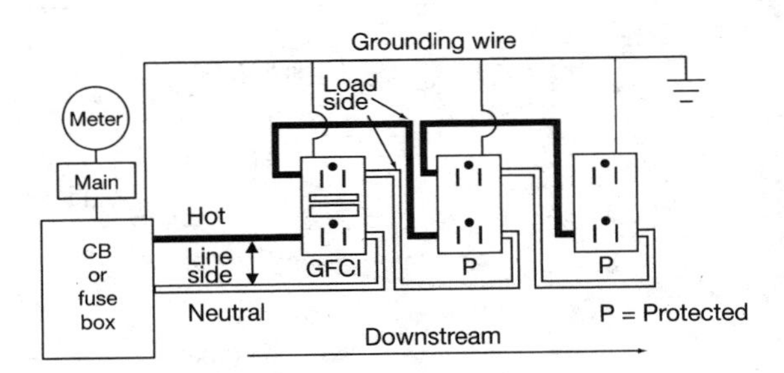

Feed-Through Installation

To protect the entire branch circuit, the GFCI must be the first receptacle from the circuit breaker or fuse box. Receptacles on the circuit downstream from the GFCI will also be protected.

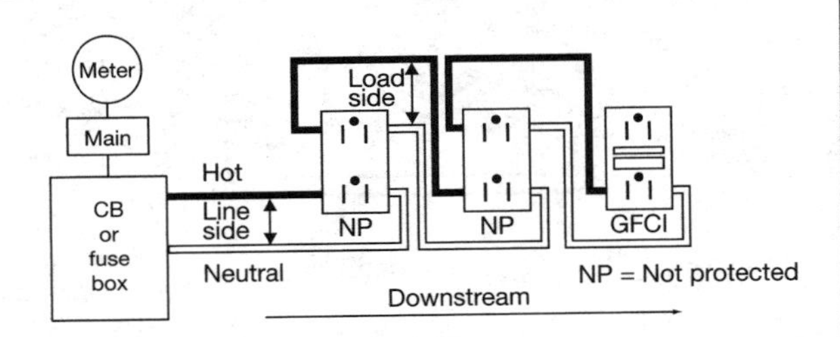

Non-Feed-Through Installation on a 2-Wire Circuit

Terminal protection can be achieved on a multi-outlet circuit by connecting the hot and neutral line conductors to the corresponding line side terminals of the GFCI. Only the GFCI receptacle will be protected.

WIRING DIAGRAM OF SPLIT-WIRED RECEPTACLES AND SWITCHED CIRCUIT

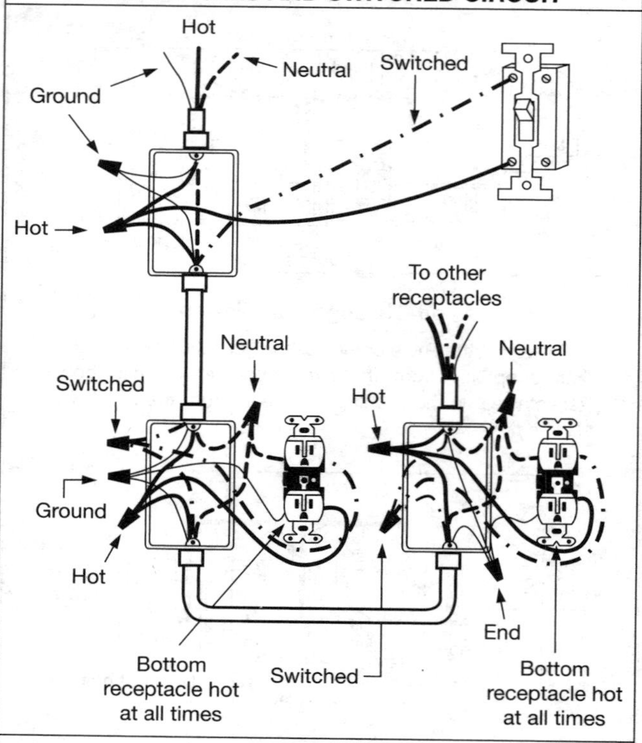

Key for Wires

Hot Wire ———
Traveler Wire ———
Neutral Wire - - - - -
Ground Wire ———
Switch Leg Wire —·—·—

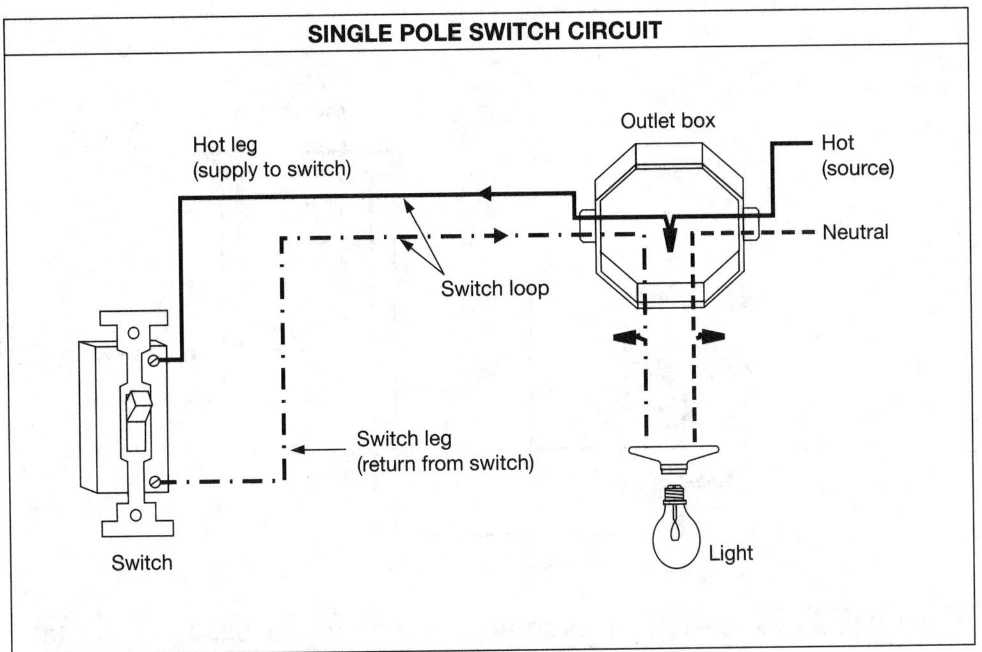
SINGLE POLE SWITCH CIRCUIT
Outlet box
Hot leg
(supply to switch)
Hot
(source)
Neutral
Switch loop
Switch leg
(return from switch)
Switch
Light

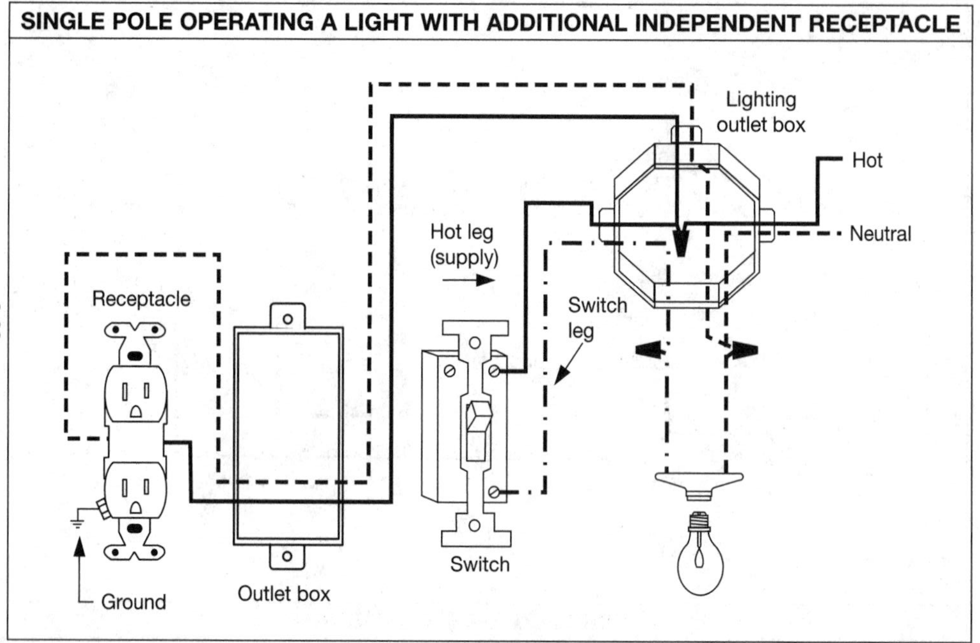
SINGLE POLE OPERATING A LIGHT WITH ADDITIONAL INDEPENDENT RECEPTACLE
Lighting
outlet box
Hot
Neutral
Hot leg
(supply)
Switch
leg
Receptacle
Outlet box
Switch
Ground

THREE-WAY SWITCH CIRCUIT OPERATING ONE OUTLET FROM TWO LOCATIONS

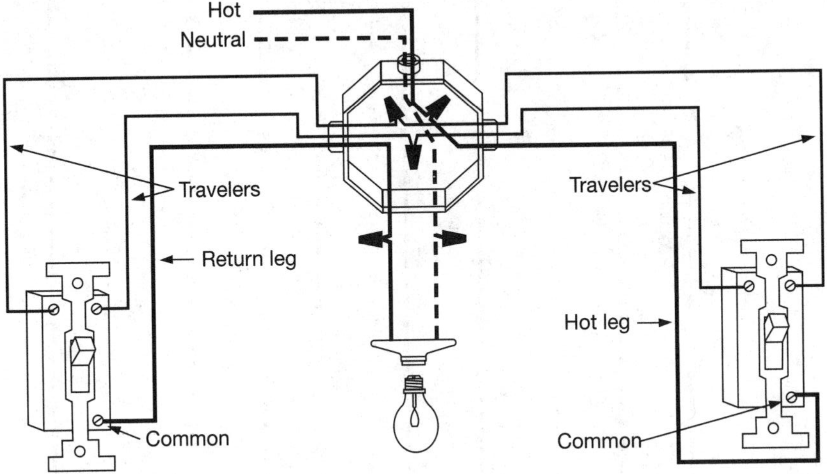

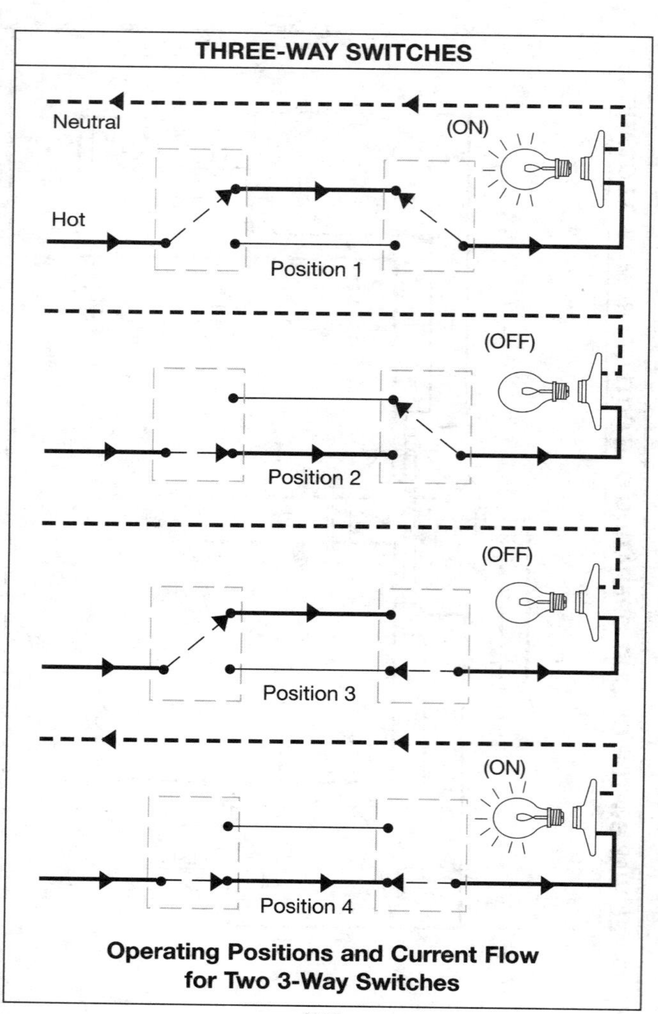

Operating Positions and Current Flow for Two 3-Way Switches

BOX FILL

Max. Number of Conductors in Outlet, Device and Junction Boxes

Box Dimension in Inches Trade Size or Type	Min. Capacity (in.³)	Maximum Number of Conductors						
		No. 18	No. 16	No. 14	No. 12	No. 10	No. 8	No. 6
4 × 1-1/4 round or octagonal	12.5	8	7	6	5	5	5	2
4 × 1-1/2 round or octagonal	15.5	10	8	7	6	6	5	3
4 × 2-1/8 round or octagonal	21.5	14	12	10	9	8	7	4
4 × 1-1/4 square	18.0	12	10	9	8	7	6	3
4 × 1-1/2 square	21.0	14	12	10	9	8	7	4
4 × 2-1/8 square	30.3	20	17	15	13	12	10	6
4-11/16 × 1-1/4 square	25.5	17	14	12	11	10	8	5
4-11/16 × 1-1/2 square	29.5	19	16	14	13	11	9	5
4-11/16 × 2-1/8 square	42.0	28	24	21	18	16	14	8
3 × 2 × 1-1/2 device	7.5	5	4	3	3	3	2	1
3 × 2 × 2 device	10.0	6	5	5	4	4	3	2
3 × 2 × 2-1/4 device	10.5	7	6	5	4	4	3	2
3 × 2 × 2-1/2 device	12.5	8	7	6	5	5	4	2
3 × 2 × 2-3/4 device	14.0	9	8	7	6	5	4	2
3 × 2 × 3-1/2 device	18.0	12	10	9	8	7	6	3
4 × 2-1/8 × 1-1/2 device	10.3	6	5	5	4	4	3	2
4 × 2-1/8 × 1-7/8 device	13.0	8	7	6	5	5	4	2
4 × 2-1/8 × 2-1/8 device	14.5	9	8	7	6	5	4	2
3-3/4 × 2 × 2-1/2 masonry box/gang	14.0	9	8	7	6	5	4	2
3-3/4 × 2 × 3-1/2 masonry box/gang	21.0	14	12	10	9	8	7	2

If one or more cable clamp is in a box, it is counted the same as the largest conductor. A loop of conductor 12 inches or more counts as two conductors.

CONDUCTOR VOLUME ALLOWANCE

Wire Size (AWG)	Volume Each (In.³)	Formula
18	1.50	V = L × W × D
16	1.75	Volume = Length times width times depth (in cubic inches)
14	2.00	
12	2.25	
10	2.50	
8	3.00	
6	5.00	

To find box size needed, add up total volume for all wires to be used. Then use the volume formula. Example: If total volume of all wires is 420 cubic inches — use an 8 × 10 × 6-inch box = 480 cubic inches.

TYPES OF CONDUCTORS

Type	Max. Temp	Application	Insulation	Outer Covering
FEP or FEPB	90°C (194°F) 200°C (392°F)	Dry and damp locations Dry locations–Special Apps	Fluorinated ethylene propylene	None or glass braid
MI	90°C (194°F) 250°C (482°F)	Dry and wet locations Special Apps	Magnesium oxide	Copper or alloy steel
MTW	60°C (140°F) 90°C (194°F)	Machine tool wiring–wet locations Machine tool wiring–dry locations	Flame-retardant, moisture, heat, and oil-resistant thermoplastic	None or nylon jacket
PAPER	85°C (185°F)	Underground service conductors	Paper	Lead sheath
PFA	90°C (194°F) 200°C (392°F)	Dry and damp locations Dry locations-Special Apps	Perfluoroalkoxy	None
PFAH	250°C (482°F)	Dry locations only	Perfluoroalkoxy	None
RHH	90°C (194°F)	Dry and damp locations		Moisture resistant, flame-retardant nonmetallic
RHW	75°C (167°F)	Dry and wet locations	Flame-retardant, moisture-resistant thermoset	Moisture-resistant, flame-retardant, nonmetallic

TYPES OF CONDUCTORS *(cont.)*

Type	Max. Temp	Application	Insulation	Outer Covering
RHW-2	90°C (194°F)	Dry and wet locations	Flame-retardant, moisture-resistant thermoset	Moisture-resistant, flame-retardant, nonmetallic
SA	90°C (194°F) 200°C (392°F)	Dry and damp locations Special Apps	Silicone rubber	Glass or braid material
SIS	90°C (194°F)	Switchboard wiring	Flame-retardant thermostat	None
TBS	90°C (194°F)	Switchboard wiring	Thermoplastic	Flame-retardant, nonmetallic
TFE	250°C (482°F)	Dry locations only	Extruded polytetrafluoroethylene	None
THHN	90°C (194°F)	Dry and damp locations	Flame-retardant, heat-resistant thermoplastic	Nylon jacket
THHW	75°C (167°F) 90°C (194°F)	Wet locations Dry locations	Flame-retardant, moisture- and heat-resistant thermoplastic	None
THW	75°C (167°F) 90°C (194°F)	Dry and wet locations Special Apps	Flame-retardant, moisture- and heat-resistant thermoplastic	None
THWN	75°C (167°F)	Dry and wet locations	Flame-retardant, moisture- and heat-resistant thermoplastic	Nylon jacket
TW	60°C (140°F)	Dry and wet locations	Flame-retardant, moisture- and heat-resistant thermoplastic	None
UF	60°C (140°F) 75°C (167°F)	Refer to NEC®	Moisture-resistant Moisture- and heat-resistant	Integral with insulation

TYPES OF CONDUCTORS *(cont.)*

Type	Max. Temp	Application	Insulation	Outer Covering
USE	75°C (167°F)	Refer to NEC®	Heat and moisture-resistant	Moisture-resistant nonmetallic
XHH	90°C (194°F)	Dry and damp locations	Flame-retardant thermoplastic	None
XHHW	90°C (194°F) 75°C (167°F)	Dry and damp locations Wet locations	Flame-retardant, moisture-resistant thermoset	None
XHHW-2	90°C (194°F)	Dry and wet locations	Flame-retardant, moisture-resistant thermoset	None
Z	90°C (194°F) 150°C (302°F)	Dry and damp locations Dry locations – Special Apps	Modified ethylene tetrafluoro ethylene	None
ZW	75°C (167°F) 90°C (194°F) 150°C (302°F)	Wet locations Dry and damp locations Dry locations – Special Apps	Modified ethylene tetrafluoro ethylene	None

CHAPTER 9
Telephone, TV, Security, Internet

TELEPHONE CIRCUIT OPERATION

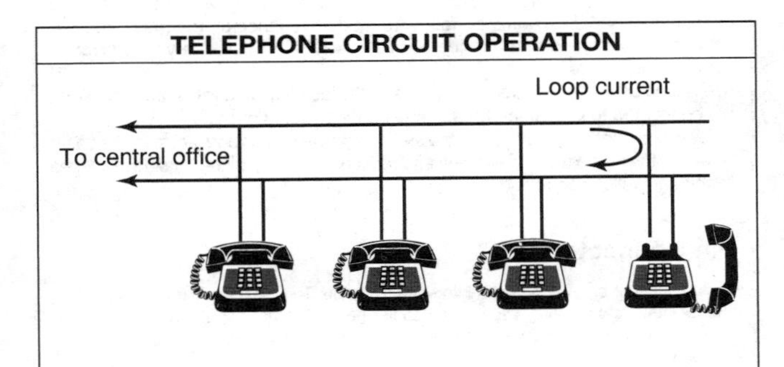

VOICE NETWORKS

Traditional Telephone Network
- Subscriber (business, residential)
- LEC (central office)
- Long Distance Carrier

Business Phone Systems
- Multiline Phone (KSUless)
- Key System
- PBX

Centrex
- Provided by the LEC
- Can coincide with a PBX or KSU
- Utilizes standard telephone sets

ISDN
- LEC must have compatible equipment
- Digital technology
- Interaction with PC network

COMMON TELEPHONE CONNECTIONS

The most common and simplest type of communication installation is the single-line telephone. The typical telephone cable (sometimes called quad cable) contains four wires, colored green, red, black, and yellow. A one-line telephone requires only two wires to operate. In almost all circumstances, green and red are the two conductors used. In a common four-wire modular connector, the green and red conductors are found in the inside positions, with the black and yellow wires in the outer positions.

As long as the two center conductors of the jack (again, always green and red) are connected to live phone lines, the telephone should operate.

Two-line phones generally use the same four-wire cables and jacks. In this case, however, the inside two wires (green and red) carry line 1, and the outside two wires (black and yellow) carry line 2.

Pin Connectors

Any number of pin-type connectors are available. The most familiar type is the RS-232 jack that is commonly used for computer ports. Another popular type of pin connector is the DB-type connector, which is the round connector commonly used for computer keyboards.

The various types of pin connectors can be used for terminating as few as five (the DB type), or more than 50 (the RS type) conductors.

50-pin *champ*-type connectors are often used with twisted-pair cables, when connecting to cross-connect equipment, patch panels, and communications equipment such as is used for networking.

Cross-Connections

Cross-connections are made at terminal *blocks*. A block is typically a rectangular, white plastic unit with metal connection points. The most common type is called a punch-down block. This is the kind that you see on the back wall of a business, where the main telephone connections are made. The wire connections are made by pushing the insulated wires into their places. When "punched" down, the connector cuts through the insulation and makes the appropriate connection.

Connections are made between punch-down blocks by using *patch cords,* which are short lengths of cable that can be terminated into the punch-down slots or that are equipped with connectors on each end.

When different systems must be connected together, cross-connects are used.

TWISTED-PAIR PLUGS AND JACKS

One of the more important factors regarding twisted-pair implementations is the cable jack or cross-connect block. These items are vital, since without the proper interface, any twisted-pair cable would be relatively useless. In the twisted-pair arena, there are three major types of twisted-pair jacks:

- RJ-type connectors (phone plugs)
- Pin-connector
- Genderless connectors (IBM sexless data connectors)

The RJ-type (registered jack) name generally refers to the standard format used for most telephone jacks. The term pin-connector refers to twisted pair connectors, such as the RS-232 connector, which provide connection through male and female pin receptacles. Genderless connectors are connectors in which no separate male or female component is present; each component can plug into any other similar component.

STANDARD PHONE JACKS

The standard phone jack is specified by a variety of different names, such as RJ and RG, which refer to their physical and electrical characteristics. These jacks consist of a male and a female component. The male component snaps into the female receptacle. The important point to note, however, is the number of conductors each type of jack can support.

Common configurations for phone jacks include support for four, six, or eight conductors. A typical example of a four-conductor jack, supporting two twisted-pairs, would be the one used for connecting most telephone handsets to their receivers.

A common six-conductor jack, supporting three twisted-pairs, is the RJ-11 jack used to connect most telephones to the telephone company or PBX systems. An example of an eight-conductor jack is the R-45 jack, which is intended for use under the ISDN system as the user-site interface for ISDN terminals.

For building wiring, the six-conductor and eight-conductor jacks are popular, with the eight-conductor jack increasing in popularity as more corporations install twisted-pair in four-pair bundles for both voice and data. The eight-conductor jack, in addition to being used for ISDN, is also specified by several other popular applications, such as the new IEEE 802.3 10 BaseT standard for Ethernet over twisted-pair.

These types of jacks are often keyed, so that the wrong type of plug cannot be inserted into the jack. There are two kinds of keying—side keying and shift keying.

Side keying uses a piece of plastic that is extended to one side of the jack. This type is often used when multiple jacks are present.

Shift keying entails shifting the position of the snap connector to the left or right of the jack, rather than leaving it in its usual center position. Shift keying is more commonly used for data connectors than for voice connectors.

Note that while we say that these jacks are used for certain types of systems (data, voice, etc.), this is not any type of standard. They can be used as you please.

COLOR CODING OF CABLES

The color coding of twisted-pair cable uses a color pattern that identifies not only what conductors make up a pair but also what pair in the sequence it is, relative to other pairs within a multipair sheath. This is also used to determine which conductor in a pair is the *tip* conductor and which is the *ring* conductor. (The tip conductor is the positive conductor, and the ring conductor is the negative conductor.)

The banding scheme uses two opposing colors to represent a single pair. One color is considered the primary while the other color is considered the secondary. For example, given the primary color of white and the secondary color of blue, a single twisted-pair would consist of one cable that is white with blue bands on it. The five primary colors are white, red, black, yellow, and violet.

In multipair cables the primary color is responsible for an entire group of pairs (five pairs total). For example, the first five pairs all have the primary color of white. Each of the secondary colors, blue, orange, green, brown, and slate, are paired in a banded fashion with white. This continues through the entire primary color scheme for all four primary colors (comprising 25 individual pairs). In larger cables (50 pairs and up), each 25-pair group is wrapped in a pair of ribbons, again representing the groups of primary colors matched with their respective secondary colors. These color-coded band markings help cable technicians to quickly identify and properly terminate cable pairs.

EIA Color Code

You should note that the new EIA color code calls for the following color coding:

Pair 1—White/Blue (white with blue stripe) and Blue
Pair 2—White/Orange and Orange
Pair 3—White/Green and Green
Pair 4—White/Brown and Brown

CATEGORY CABLING

Category 1 cable is the old standard type of telephone cable with four conductors colored green, red, black, and yellow. Also called quad cable.
Category 2 Obsolete.
Category 3 cable is used for digital voice and data transmission rates up to 10 Mbit/s (Megabits per second). Common types of data transmission over this communications cable would be UTP Token Ring (4 Mbit/s) and 10Base-T (10 Mbit/s).
Category 4 Obsolete.
Category 5 cable is used for sending voice and data at speeds up to 100 Mbit/s (megabits per second).
Category 6 cable is used for sending data at speeds up to 200 or 250 Mbit/s (megabits per second).

STANDARD TELECOM COLOR CODING

Pair #	Tip (+) Color	Ring (–) Color
1	White	Blue
2	White	Orange
3	White	Green
4	White	Brown
5	White	Slate
6	Red	Blue
7	Red	Orange
8	Red	Green
9	Red	Brown
10	Red	Slate
11	Black	Blue
12	Black	Orange
13	Black	Green
14	Black	Brown
15	Black	Slate
16	Yellow	Blue
17	Yellow	Orange
18	Yellow	Green
19	Yellow	Brown
20	Yellow	Slate
21	Violet	Blue
22	Violet	Orange
23	Violet	Green
24	Violet	Brown
25	Violet	Slate

66 BLOCK WIRING AND CABLE COLOR CODING

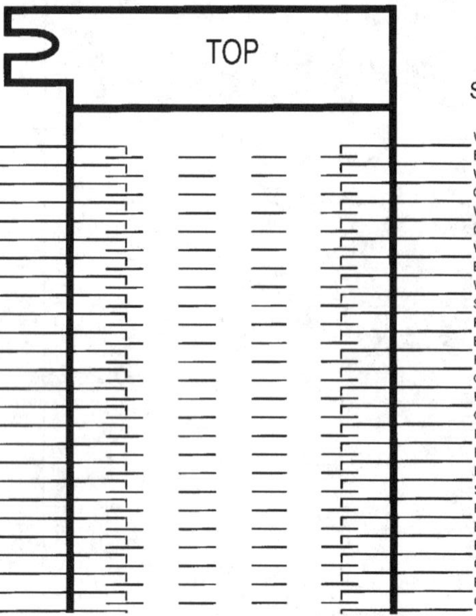

TOP

PAIR CODE			SIDE #1	SIDE #2
Pair 1	Tip	26	White/Blue	White/Blue
	Ring	1	Blue/White	Blue/White
Pair 2	Tip	27	White/Orange	White/Orange
	Ring	2	Orange/White	Orange/White
Pair 3	Tip	28	White/Green	White/Green
	Ring	3	Green/White	Green/White
Pair 4	Tip	29	White/Brown	White/Brown
	Ring	4	Brown/White	Brown/White
Pair 5	Tip	30	White/Slate	White/Slate
	Ring	5	Slate/White	Slate/White
Pair 6	Tip	31	Red/Blue	Red/Blue
	Ring	6	Blue/Red	Blue/Red
Pair 7	Tip	32	Red/Orange	Red/Orange
	Ring	7	Orange/Red	Orange/Red
Pair 8	Tip	33	Red/Green	Red/Green
	Ring	8	Green/Red	Green/Red
Pair 9	Tip	34	Red/Brown	Red/Brown
	Ring	9	Brown/Red	Brown/Red
Pair 10	Tip	35	Red/Slate	Red/Slate
	Ring	10	Slate/Red	Slate/Red
Pair 11	Tip	36	Black/Blue	Black/Blue
	Ring	11	Blue/Black	Blue/Black
Pair 12	Tip	37	Black/Orange	Black/Orange
	Ring	12	Orange/Black	Orange/Black
Pair 13	Tip	38	Black/Green	Black/Green
	Ring	13	Green/Black	Green/Black

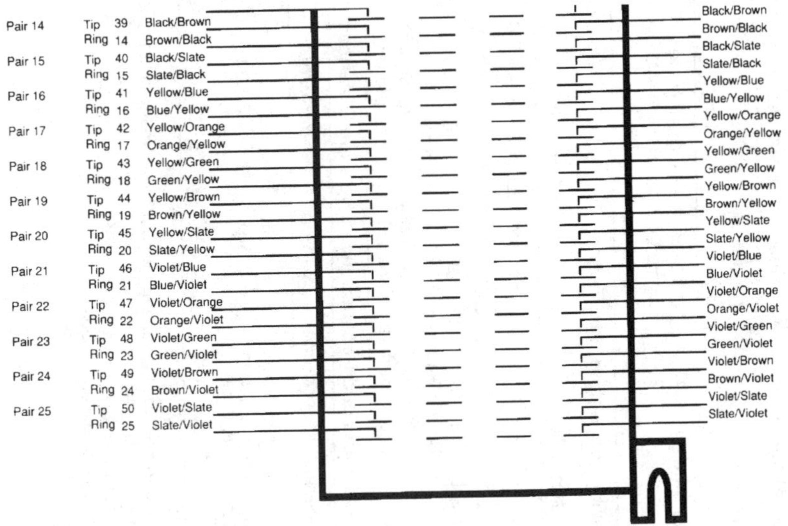

Pair	Lead	Pin	Color	Color
Pair 14	Tip	39	Black/Brown	Black/Brown
	Ring	14	Brown/Black	Brown/Black
Pair 15	Tip	40	Black/Slate	Black/Slate
	Ring	15	Slate/Black	Slate/Black
Pair 16	Tip	41	Yellow/Blue	Yellow/Blue
	Ring	16	Blue/Yellow	Blue/Yellow
Pair 17	Tip	42	Yellow/Orange	Yellow/Orange
	Ring	17	Orange/Yellow	Orange/Yellow
Pair 18	Tip	43	Yellow/Green	Yellow/Green
	Ring	18	Green/Yellow	Green/Yellow
Pair 19	Tip	44	Yellow/Brown	Yellow/Brown
	Ring	19	Brown/Yellow	Brown/Yellow
Pair 20	Tip	45	Yellow/Slate	Yellow/Slate
	Ring	20	Slate/Yellow	Slate/Yellow
Pair 21	Tip	46	Violet/Blue	Violet/Blue
	Ring	21	Blue/Violet	Blue/Violet
Pair 22	Tip	47	Violet/Orange	Violet/Orange
	Ring	22	Orange/Violet	Orange/Violet
Pair 23	Tip	48	Violet/Green	Violet/Green
	Ring	23	Green/Violet	Green/Violet
Pair 24	Tip	49	Violet/Brown	Violet/Brown
	Ring	24	Brown/Violet	Brown/Violet
Pair 25	Tip	50	Violet/Slate	Violet/Slate
	Ring	25	Slate/Violet	Slate/Violet

TELEPHONE CONNECTIONS

Typical Inside Wire

Type of Wire	Pair No.	Pair Color Matches	
2-pair Wire	1 2	Green Black	Red Yellow
3-pair Wire	1 2 3	White/Blue White/Orange White/Green	Blue/White Orange/White Green/White

Inside Wire Connecting Terminations

Wire Color		Wire Function	
2-pair wire	3-pair wire	Service w/o Dial Light	Service with Dial Light
Green Red Black Yellow	White/Blue Blue/White White/Orange Orange/White	Tip Ring Not Used Ground	Tip Ring Transformer Transformer

Typical Fasteners and Recommended Spacing Distances

Fasteners	Horizontal	Vertical	From Corner
Wire clamp	16 in.	16 in.	2 in.
Staples(wire)	7.5 in.	7.5 in.	2 in.
Bridle Rings*	4 ft.		2–8.5 in.*
Drive Rings**	4 ft.	8 ft.	2–8.5 in.*

*When changing direction the fasteners should be spaced to hold the wire at approximately a 45-degree angle.

**To avoid possible injury, do not use drive rings below a 6-foot clearance level; instead, use bridle rings.

SEPARATION AND PHYSICAL PROTECTION FOR PREMISES WIRING

This table applies only to telephone wiring from the Network Interface or other telephone company-provided modular jacks to telephone equipment. Minimum separations between telephone wiring whether located inside or attached to the outside of buildings and other types of wiring involved are as follows. Separations apply to crossing and to parallel runs (minimum separations).

Types of Wire Involved		Minimum Separations	Wire Crossing Alternatives
Electric Supply	Bare light or power wire of any voltage	5 ft.	No alternative
	Open wiring not over 300 volts	2 in.	See Note 1.
	Wires in conduit or in armored or nonmetallic sheath cable, or power ground wires	None	N/A
Radio & TV	Antenna lead-in and ground wires	4 in.	See Note 1.
Signal or Control Wires	Open wiring or wires in conduit or cable	None	N/A
Comm. Wires	Community Television systems coaxial cables with grounded shielding	None	N/A
Telephone Drop Wire	Using fused protectors Using fuseless protector or where no protector wiring from transformer	2 in.	See Note 1. None
Sign	Neon Signs and associated wiring from transformer		6 in. No alternative
Lightning Systems	Lightning rods and wires		6 ft.

NOTE 1: If minimum separations cannot be obtained, additional protection of a plastic tube, wire guard, or two layers of vinyl tape extending 2 inches beyond each side of object being crossed must be provided.

COMMON WIRING CONFIGURATIONS

USOC wiring is available for 1-, 2-, 3-, or 4-pair systems. Pair 1 occupies the center conductors, pair 2 occupies the next two contacts out, etc. One advantage to this scheme is that a 6-position plug configured with 1, 2, or 3 pairs can be inserted into an 8-position jack and maintain pair continuity. A note of warning though: Pins 1 and 8 on the jack may become damaged from this practice. A disadvantage is the poor transmission performance associated with this type of pair sequence.

USOC 4-Pair

Pair ID	PIN #
T1	5
R1	4
T2	3
R2	6
T3	2
R3	7
T4	1
R4	8

USOC 1-, 2-, or 3-Pair

Pair ID	PIN #
T1	4
R1	3
T2	2
R2	5
T3	1
R3	6

Ethernet 10BaseT

Ethernet 10BaseT wiring specifies an 8-position jack but uses only two pairs. These are pairs two and three of TIA schemes.

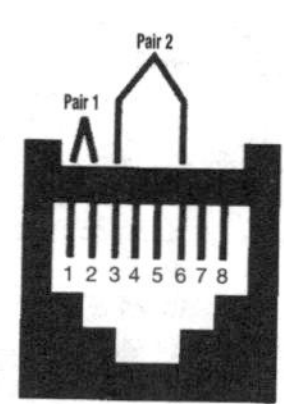

Pair ID	PIN #
T1	1
R1	2
T2	3
R2	6

UNDERSTANDING DECIBELS

In data cabling, most energy and power levels, losses, or attenuations are expressed in decibels rather than in watts. The reason is simple: Transmission calculations and measurements are almost always made as *comparisons* against a reference—received power compared to emitted power, energy in versus energy out (energy lost in a connection), etc.

- Generally, energy levels (emission, reception, etc.) are expressed in *dBm.* This signifies that the reference level of 0 dBm corresponds to 1 mW of power.
- Generally, power losses or gains (attenuation in a cable, loss in a connector, etc.) are expressed in dB.
- The unit dBμ is used for very low levels.
- Decibel measurement works as follows—a difference of 3 dB equals a doubling or halving of power.
- A 3 dB gain in power means that the optical power has been doubled. A 6 dB gain means that the power has been doubled, and doubled again, equaling four times the original power. A 3 dB loss of power means that the power has been cut in half. A 6 dB loss means that the power has been cut in half, then cut in half again, equaling one fourth of the original power.
- A loss of 3 dB in power is equivalent to a 50% loss. For example, 1 milliwatt of power in, and .5 milliwatt of power out.
- A 6 dB loss would equal a 75% loss (1 milliwatt in, .25 milliwatt out).

EIA 570 RESIDENTIAL NETWORKS

Recognized Cabling by Grade

Cabling	Grade 1	Grade 2
4-pair UTP	1 Cable per outlet	2 Cables per outlet
	Cat 3 minimum,	Cat 5 Minimum,
	Cat 5 recommended	Cat 5e recommended
75-ohm	1 Cable per outlet	2 Cables per outlet
Coax	Series 6	Series 6
Fiber	Not recommended	Optional
		50 or 62.5 Micron multimode

Space Allocation for the Distribution Device

Number of Outlets	Grade 1 (height by width)	Grade 2 (height by width)
1 to 8	24" by 16"	36" by 32"
9 to 16	36" by 16"	36" by 32"
17 to 24	48" by 16"	48" by 32"
Greater than 24	60" by 16"	60" by 32"

TYPICAL RESIDENTIAL SERVICES SUPPORTED BY GRADE

Service	Grade 1	Grade 2
Telephone	X	X
Television	X	X
Data	X	X
Multimedia		X

Recognized Residential Cabling by Grade

Cabling	Grade 1	Grade 2
4-Pair UTP	Category 3; Category 5 cable recommended	Category 5; Category 5e cable recommended
75-ohm coax	X	X
Fiber		X (optional)

T568A EIGHT-POSITION PIN-PAIR ASSIGNMENT

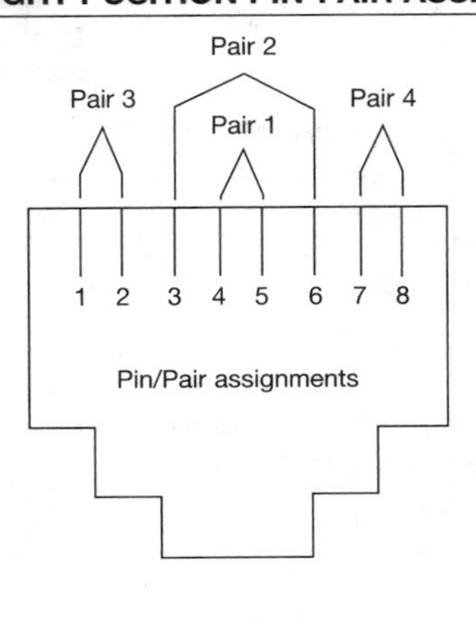

TYPICAL EIA 570 CABLING SYSTEM

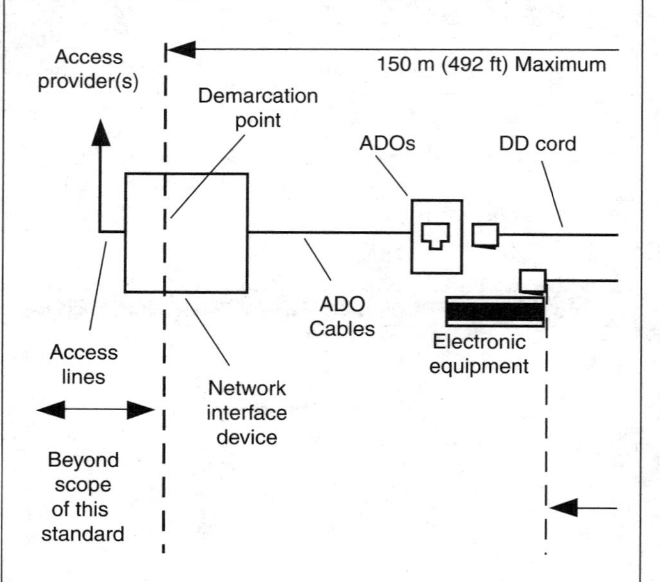

ADO—Auxiliary disconnect outlet
DD—Distribution device

FOR SINGLE RESIDENTIAL NETWORKS

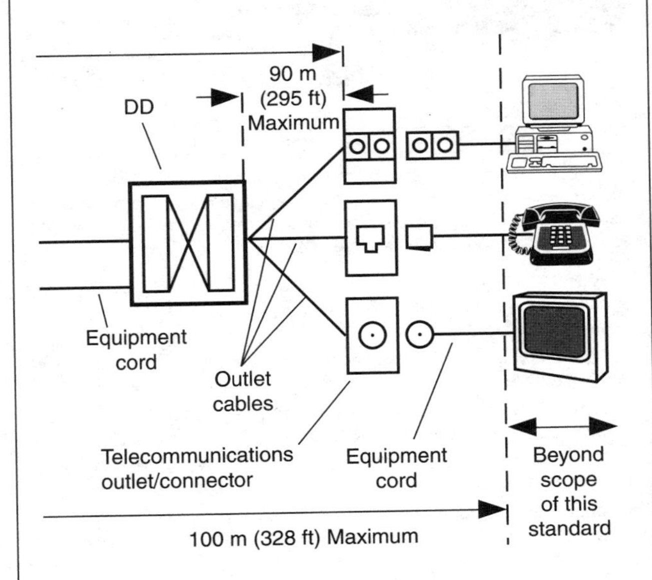

ADO—Auxiliary disconnect outlet
DD—Distribution device

COAXIAL CABLE CONSTRUCTION

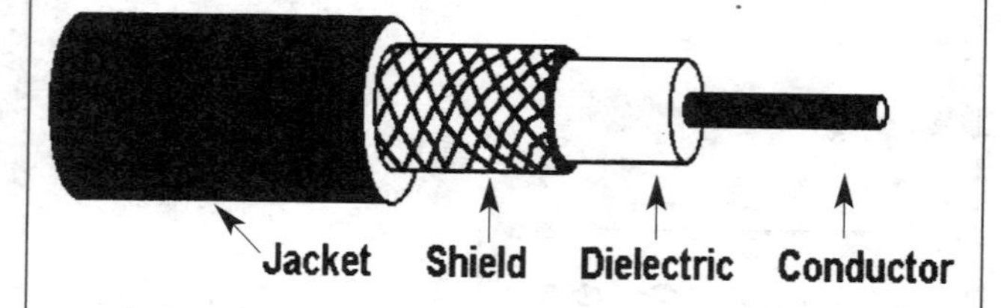

COAXIAL CABLE CLASSIFICATIONS

The NEC goes into great detail on designated cable types. Many of these requirements apply more to the cable manufacturer than to the installer. Nonetheless, the proper cable type must be used for the installation. The NEC designations and their uses are as follow:

- **Type CATVP.** CATVP is plenum cable (hence the "P" designation), and may be used in plenums, ducts, or other spaces for environmental air.
- **Type CATVR.** CATVR is riser cable, and is made suitable (extremely fire resistant) to run installations in shafts or from floor to floor in buildings.
- **Type CATV.** CATV is general use cable. It can be used in almost any location, except for risers and plenums.
- **Type CATVX.** CATVX is a limited-use cable, and is allowed only in dwellings and in raceways.
- Many coax cables are multiple-rated. In other words, their jacket is tested and suitable for several different applications. In such cases, they will be stamped with all of the applicable markings, such as CATV or CATVR.
- All coaxial cable types start with "CATV". Although you may think that the term refers to cable television, it does not. It refers to Community Antenna Television.
- Trade designations generally refer to the cable's electrical characteristics; specifically, the impedance of the cable. This is why different cable types (RG59U, RG58U, etc.) should not be mixed, even though they appear to be virtually identical—they have differing levels of impedance, and mixing them may degrade system performance.

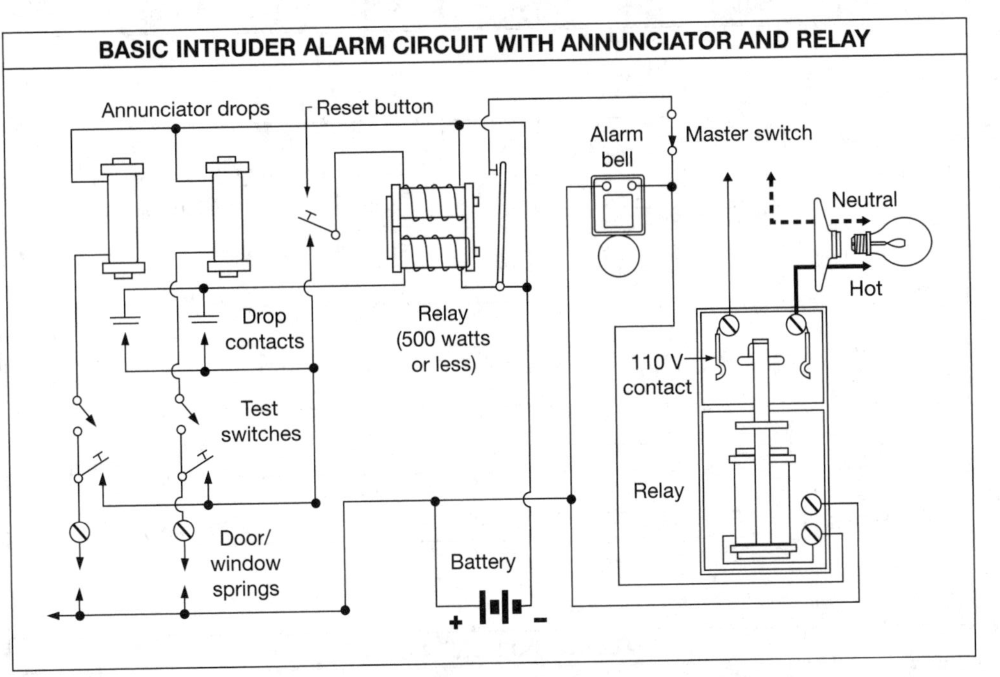
BASIC INTRUDER ALARM CIRCUIT WITH ANNUNCIATOR AND RELAY
Annunciator drops
Reset button
Alarm
bell
Master switch
Neutral
Hot
Drop
contacts
Relay
(500 watts
or less)
110 V
contact
Test
switches
Relay
Door/
window
springs
Battery
+
–

BASIC ALARM CIRCUITRY

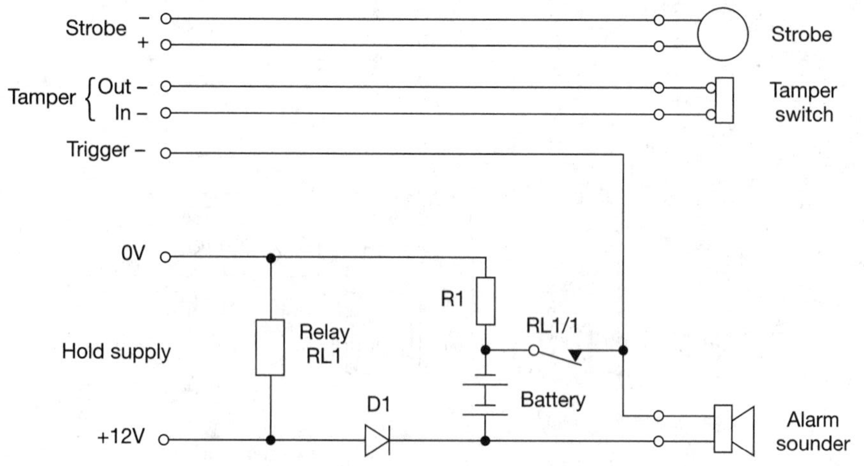

Strobe –
Strobe +
Tamper Out –
Tamper In –
Trigger –
0V
Hold supply
+12V
Relay RL1
R1
D1
RL1/1
Battery
Strobe
Tamper switch
Alarm sounder

LATCHING RELAY CIRCUIT

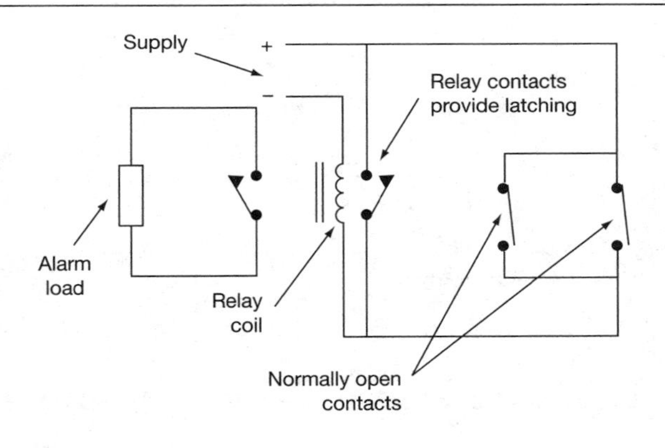

NORMALLY OPEN SINGLE-LOOP CONFIGURATION

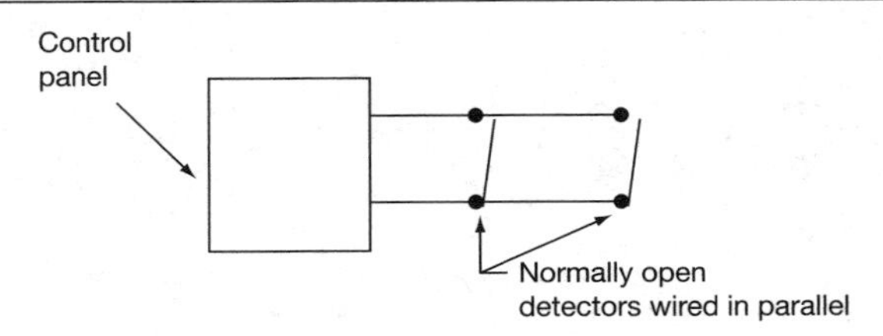

No alarm condition shown. If either sensor closes, an alarm condition exists.

BELL CIRCUIT FOR ONE DOOR

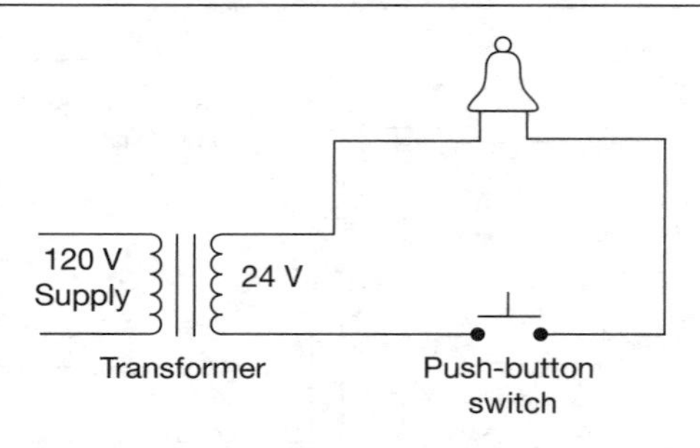

BELL CONTROLLED FROM SEVERAL LOCATIONS

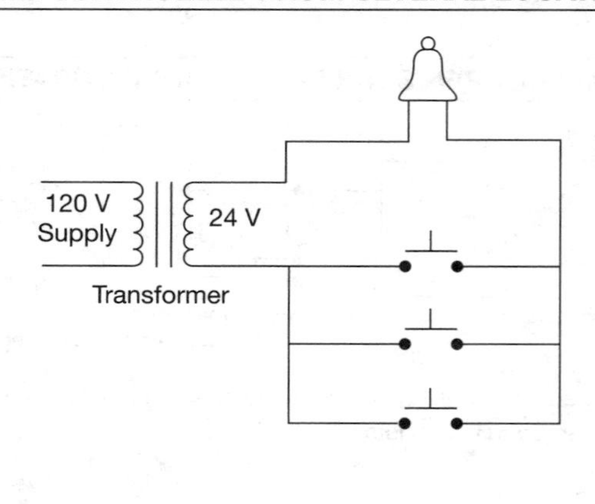

CENTRAL CHIME FOR TWO DOORS

Door 1
Push button

Door 2
Push button

Neutral

24 V

120 V

THREE-TONE DOOR SIGNAL SYSTEMS

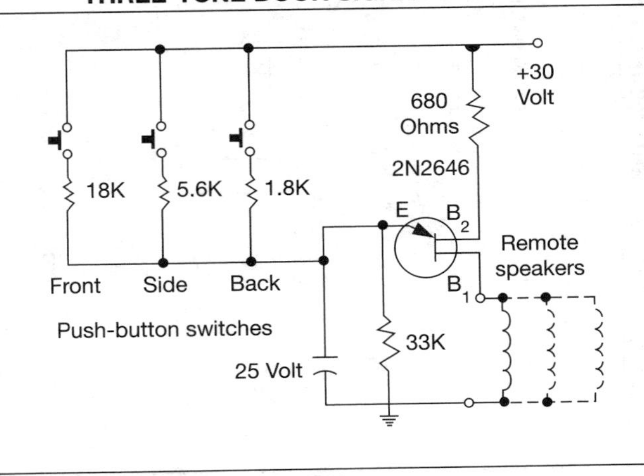

GROUNDING AN OUTDOOR ANTENNA

CHAPTER 10
Materials and Tools

COMMON NAILS				
Pennyweight	Length (inch)	Body Diameter (inch)	Head Diameter (inch)	Weight (number per pound)
Aluminum				
4d	1-½	0.099	0.250	830
6d	2	0.120	0.266	430
8d	2-½	0.148	0.281	220
10d	3	0.162	0.312	170
16d	3-½	0.177	0.344	120
20d	4	0.199	0.406	78

CASING NAILS				
Pennyweight	Length (inch)	Body Diameter (inch)	Head Diameter (inch)	Weight (number per pound)
2d	1	0.067	0.099	1,090
3d	1-¼	0.076	0.113	650
4d	1-½	0.080	0.120	490
5d	1-¾	0.080	0.120	415
6d	2	0.099	0.142	245
7d	2-¼	0.099	0.142	215
8d	2-½	0.113	0.155	150
9d	2-¾	0.113	0.155	135
10d	3	0.128	0.170	95
12d	3-¼	0.128	0.170	90
16d	3-½	0.135	0.177	75

CONCRETE NAILS

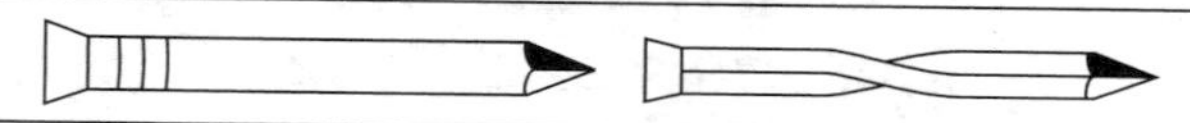

Length (inch)	Body Diameter (inch)	Head Diameter (inch)	Weight (number per pound)
Smooth Shank			
½	0.148	0.312	450
⅝	0.148	0.312	350
¾	0.148	0.312	290
⅞	0.148	0.312	250
1	0.148	0.312	210
Deformed Shank			
¾	0.181	0.284	240
1	0.181	0.284	204
1-½	0.181	0.284	118
1-¾	0.181	0.284	112
2	0.181	0.284	93
2-½	0.181	0.284	68
2-¾	0.181	0.284	60
3	0.181	0.284	52

DOUBLE-HEADED NAILS

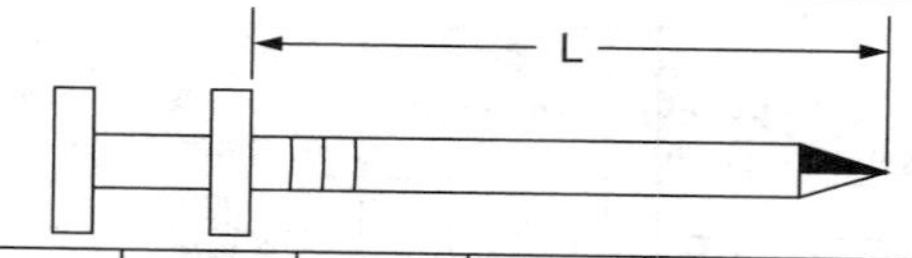

Pennyweight	Total Length (inch)	Body Length (inch)	Head Length (inch)	Body Diameter (inch)	Weight (number per pound)
6d	2	1-¾	¼	0.113	180
8d	2-½	2-¼	¼	0.131	90
10d	3-1/16	2-¾	5/16	0.148	59
18d	3-⅜	3	⅜	0.162	45
20d	3-⅞	3-½	⅜	0.192	28
30d	4-7/16	4	7/16	0.207	22

FINISH NAILS

Pennyweight	Length (inch)	Body Diameter (inch)	Head Diameter (inch)	Weight (number per pound)
2d	1	0.058	0.086	1,470
3d	1-¼	0.067	0.099	880
4d	1-½	0.072	0.106	630
5d	1-¾	0.072	0.106	530
6d	2	0.092	0.135	290
7d	2-¼	0.092	0.135	250
8d	2-½	0.099	0.142	190
9d	2-¾	0.099	0.142	180
10d	3	0.113	0.155	120
12d	3-¼	0.113	0.155	110
18d	3-½	0.120	0.162	93
20d	4	0.135	0.177	85

FLOORING NAILS

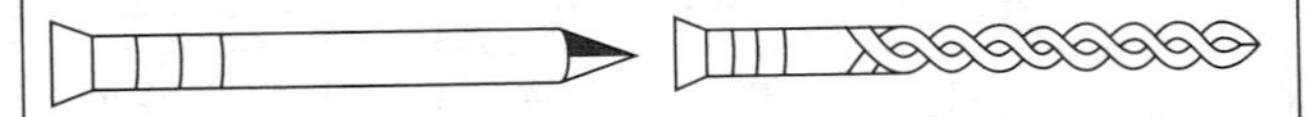

Pennyweight	Length (inch)	Body Diameter (inch)	Head Diameter (inch)	Weight (number per pound)
Smooth and Deformed				
2d	1	0.072	0.141	840
3d	1-¼	0.072	0.141	700
4d	1-½	0.080	0.156	430
4d	1-½	0.092	0.156	370
5d	1-¾	0.092	0.156	310
6d	2	0.113	0.203	180
7d	2-¼	0.113	0.203	150
8d	2-½	0.135	0.177	100
8d	2-½	0.113	0.203	110
10d	3	0.135	0.250	82
12d	3-¼	0.135	0.250	75
16d	3-½	0.148	0.281	58

GYPSUM WALLBOARD NAILS

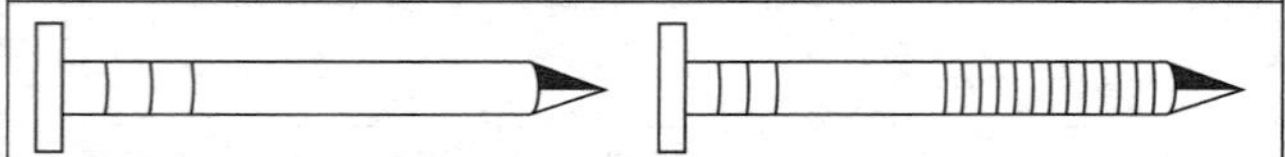

Length (inch)	Body Diameter (inch)	Head Diameter (inch)	Weight (number per pound)
Smooth Shank			
1-⅛	0.092	0.297	470
1-¼	0.092	0.297	420
1-¾	0.092	0.375	290
Deformed Shank			
1-⅛	0.099	0.250	380
1-¼	0.099	0.250	340
1-½	0.099	0.250	290
1-⅝	0.099	0.250	270

SIDING NAILS

Length (inch)	Body Diameter (inch)	Head Diameter (inch)	Weight (number per pound)
Aluminum, Flat Head			
1-½	0.113	0.312	680
2	0.113	0.219	490
2-½	0.135	0.219	290
Aluminum, Casing Head			
1-⅞	0.106	0.141	600
2-⅛	0.113	0.141	470
2-⅜	0.128	0.158	320
2-⅝	0.148	0.189	200
Aluminum, Countersunk Head			
1-⅞	0.106	0.266	600
2-⅛	0.113	0.286	470
2-⅜	0.128	0.297	320
2-⅝	0.148	0.312	200

JOIST HANGER NAILS

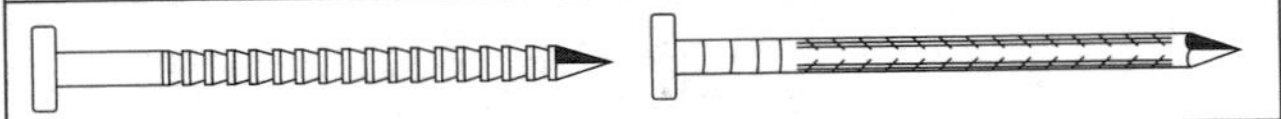

Pennyweight	Length (inch)	Body Diameter (inch)	Steel Type	Weight (number per pound)
Smooth Shank				
8d	1-¼	—	B	230
8d	1-½	0.131	G	175
8d	1-½	0.131	G	152
8d	1-½	0.131	S	152
8d	1-½	0.131	S	143
8d	2-½	0.131	S	94
10d	1-½	0.148	G	128
10d	1-½	—	B	128
10d	1-½	0.148	G	119
10d	1-½	0.148	S	122
10d	1-½	0.148	S	112
10d	3	0.148	S	67
16d	2-½	0.162	G	66
16d	2-½	—	B	65
16d	2-½	0.162	S	44
16d	2-½	0.162	B	63
Deformed Shank				
10d	3	0.132	B	—
20d	1-¾	0.192	G	65
20d	1-¾	0.192	B	63
20d	1-¾	—	B	63
20d	2-⅛	—	B	50
20d	2-½	0.192	G	46
—	2-½	0.250	B	27
—	3	0.250	B	27
—	3	0.250	G	22

B = Bright, G = Galvanized, S = Stainless

UNDERLAYMENT NAILS

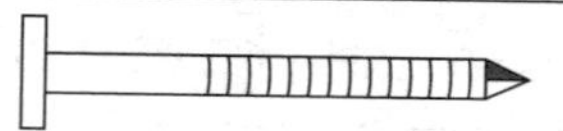

Length (inch)	Body Diameter (inch)	Head Diameter (inch)	Weight (number per pound)
1	0.080	0.188	—
1-¼	0.080	0.188	600
1-¼	0.099	0.250	400
1-⅜	0.080	0.188	540
1-⅜	0.099	0.250	360
1-½	0.080	0.188	500
1-½	0.099	0.250	330
1-⅝	0.099	0.250	300
1-¾	0.099	0.250	280
1-⅞	0.108	0.266	170
2-⅛	0.108	0.266	170
2-⅜	0.113	0.297	140

MASONRY DRIVE NAILS

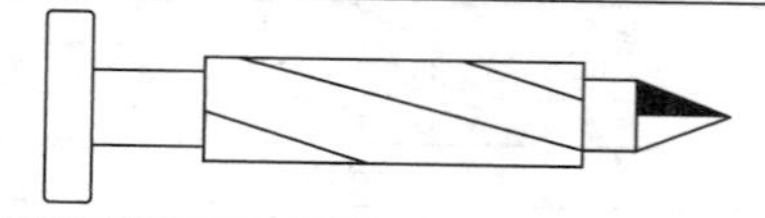

Length (inch)	Head Length (inch)	Thread Diameter (inch)
¾	3/32	0.125
¾	⅛	0.156
1	5/32	0.188
1-¼	3/16	0.215
1-½	¼	0.258
2	5/16	0.330

MASONRY NAILS

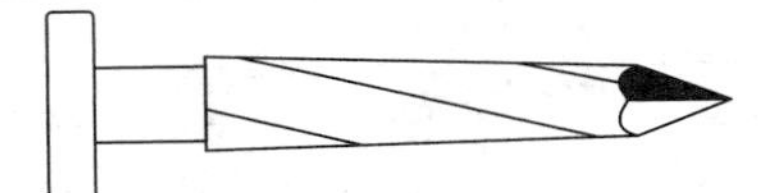

Length (inch)	Body Diameter (inch)	Head Diameter (inch)	Weight (number per pound)
Standard			
½	0.148	0.312	340
¾	0.148	0.312	280
1	0.148	0.312	170
1-¼	0.148	0.312	140
1-½	0.148	0.312	130
1-¾	0.148	0.312	110
2	0.148	0.312	98
2-¼	0.148	0.312	84
2-½	0.148	0.312	78
2-¾	0.148	0.312	70
3	0.148	0.312	67
3-¼	0.148	0.312	60
3-½	0.162	0.344	48
3-¾	0.162	0.344	45
4	0.177	0.375	35
Heavy			
1	0.250	0.582	53
1-¼	0.250	0.582	47
1-½	0.250	0.582	43
1-¾	0.250	0.582	39
2	0.250	0.582	34
2-½	0.250	0.582	27
3-½	0.250	0.582	19
3	0.250	0.582	24

ROOFING NAILS

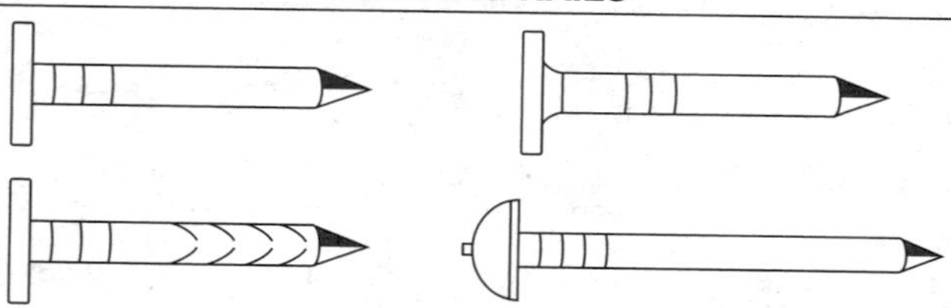

Length (inch)	Body Diameter (inch)	Head Diameter (inch)	Weight (number per pound)
Aluminum			
¾	0.120	0.438	940
¾	0.135	0.438	750
⅞	0.120	0.438	830
⅞	0.135	0.438	680
1	0.120	0.438	700
1	0.135	0.438	600
1	0.135	0.438	580
1-¼	0.120	0.438	620
1-¼	0.135	0.438	490
1-½	0.120	0.438	520
1-½	0.135	0.438	420
1-¾	0.135	0.438	370
2	0.135	0.438	340
2-½	0.145	0.438	230
Steel			
¾	0.106	0.375	460
¾	0.120	0.438	340
¾	0.135	0.469	270
¾	0.142	0.484	240
¾	0.148	0.500	220
¾	0.162	0.500	200
⅞	0.106	0.375	–
⅞	0.120	0.438	300
⅞	0.120	0.500	250
⅞	0.135	0.469	240
⅞	0.142	0.484	210

ROOFING NAILS *(cont.)*			
Length (inch)	Body Diameter (inch)	Head Diameter (inch)	Weight (number per pound)
	Steel *(cont.)*		
7/8	0.148	0.500	190
7/8	0.162	0.500	170
1	0.106	0.281	380
1	0.106	0.281	360
1	0.120	0.438	270
1	0.120	0.500	220
1	0.135	0.469	210
1	0.142	0.484	190
1	0.148	0.500	170
1	0.162	0.500	150
1-1/8	0.106	0.375	320
1-1/8	0.120	0.438	240
1-1/8	0.135	0.469	190
1-1/8	0.142	0.484	170
1-1/8	0.148	0.500	160
1-1/8	0.162	0.500	140
1-1/4	0.106	0.375	300
1-1/4	0.120	0.312	240
1-1/4	0.120	0.438	220
1-1/4	0.120	0.500	—
1-1/4	0.135	0.469	180
1-1/4	0.142	0.484	160
1-1/4	0.148	0.500	140
1-1/4	0.162	0.500	120
1-1/2	0.106	0.375	—
1-1/2	0.120	0.438	180
1-1/2	0.120	0.500	160
1-1/2	0.135	0.489	150
1-1/2	0.142	0.484	130
1-1/2	0.148	0.500	120
1-1/2	0.162	0.500	110
1-3/4	0.106	0.375	220
1-3/4	0.120	0.438	160
1-3/4	0.120	0.500	140
1-3/4	0.135	0.469	130

ROOFING NAILS *(cont.)*			
Length (inch)	**Body Diameter (inch)**	**Head Diameter (inch)**	**Weight (number per pound)**
Steel *(cont.)*			
1-3/4	0.142	0.484	120
1-3/4	0.148	0.500	110
1-3/4	0.162	0.500	92
Steel, Copper-clad			
1	0.120	0.375	280
1-1/4	0.120	0.375	220
1-1/2	0.120	0.375	190
1-3/4	0.120	0.375	160
2	0.120	0.375	140
2-1/4	0.120	0.375	130
Steel, Leak-resistant, Convex Head			
1-3/4	0.135	0.500	110
2	0.135	0.500	98
Steel, Reinforced Head			
3/4	0.106	0.625	190
3/4	0.120	0.625	170
7/8	0.106	0.625	180
7/8	0.120	0.625	160
1	0.106	0.625	170
1	0.120	0.625	150
1-1/8	0.106	0.625	170
1-1/8	0.120	0.625	140
1-1/4	0.106	0.625	160
1-1/4	0.106	0.625	140
Steel, Mechanically Deformed Shank			
1/2	0.106	—	130
5/8	0.106	—	120
3/4	0.106	—	115
7/8	0.106	—	110
1	0.106	—	110
1-1/8	0.106	—	110
1-1/4	0.106	—	100

ROOFING NAILS *(cont.)*			
Length (inch)	**Body Diameter (inch)**	**Head Diameter (inch)**	**Weight (number per pound)**
Steel, Mechanically Deformed Shank *(cont.)*			
1-½	0.106–0.120	–	96–84
1-¾	0.106–0.120	–	94–85
2	0.106–0.120	–	90–74
2-½	0.106–0.120	–	80–61
3	0.106	–	70
Steel, Cast Lead Head, Smooth Shank			
1-½	0.148	–	98
1-¾	0.148	–	87
2	0.148	–	79
Steel, Cast Lead Head, Barbed Shank			
1-½	0.135	–	110
1-¾	0.135	–	110
2	0.135	–	93
Aluminum, Neoprene Washer, Smooth Shank			
1-¾	0.135	0.438	320
2	0.135	0.438	280
2-¼	0.135	0.438	240
2-½	0.135	0.438	210
Aluminum, Neoprene Washer, Deformed Shank			
1-¾	0.145	0.438	290
2	0.145	0.438	280
2-¼	0.145	0.438	230
2-½	0.145	0.438	210

SHINGLE NAILS

Length (inch)	Body Diameter (inch)	Head Diameter (inch)	Weight (number per pound)
Aluminum, Deformed Shank			
1-¼	0.101	0.191	1,060
1-½	0.101	0.191	860
1-¾	0.105	0.191	720
2	0.105	0.191	610
2-¼	0.113	0.200	180
2-½	0.113	0.200	130
Aluminum, Smooth Shank			
⅞	0.099	0.281	1,310
1-¼	0.080	0.219	1,460
1-¼	0.099	0.281	1,010
1-¼	0.113	0.312	780
1-½	0.113	0.312	660
1-¾	0.113	0.312	610
Steel, Smooth Shank			
1-¼	0.092	0.250	410
1-⅜	0.099	0.281	310
1-½	0.108	0.281	260
Steel, Barbed Shank			
1-¼	0.113	0.406	250
1-½	0.113	0.406	210
1-¾	0.113	0.406	180
2	0.113	0.406	162

COMMON SPIKES

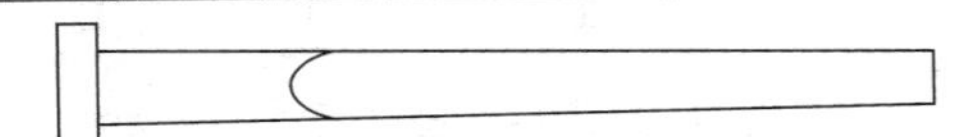

Length (inch)	Pennyweight
4	20d
4-½	30d
5	40d
5-½	50d
6	60d
7	80d
8	100d

GUTTER SPIKES

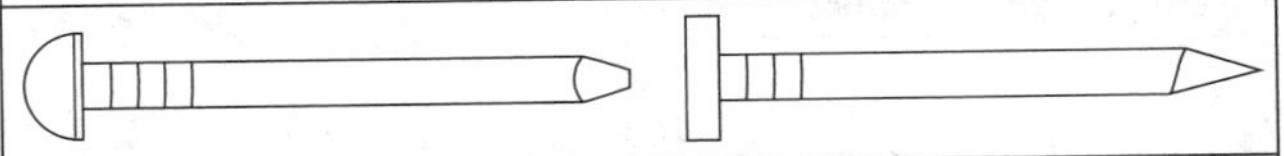

Length (inch)	Body Diameter (inch)	Head Diameter (inch)
	Flat Head	
6-½	0.250	0.562
7	0.250	0.562
8	0.250	0.562
8-½	0.250	0.562
9	0.250	0.562
10	0.250	0.562
10-½	0.250	0.562
	Oval Head	
6-½	0.250	0.531
7	0.250	0.531
8	0.250	0.531
8-½	0.250	0.531
9	0.250	0.531
10	0.250	0.531
10-½	0.250	0.531

BOX NAILS

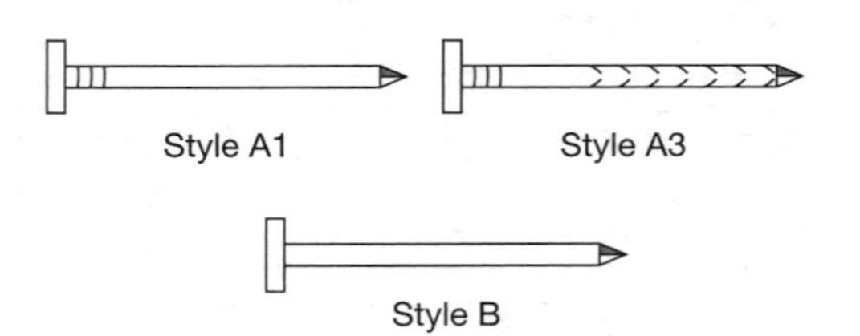

Pennyweight	Length (inch)	Body Diameter (inch)	Head Diameter (inch)	Weight (number per pound)
Style A				
2d	1	0.067	0.188	940
3d	1-¼	0.076	0.219	590
4d	1-½	0.080	0.219	450
5d	1-¾	0.080	0.219	390
6d	2	0.099	0.266	220
7d	2-¼	0.099	0.266	200
8d	2-½	0.113	0.297	140
9d	2-¾	0.113	0.297	120
10d	3	0.128	0.312	90
12d	3-¼	0.128	0.312	83
16d	3-½	0.135	0.344	69
20d	4	0.148	0.375	50
30d	4-½	0.148	0.375	45
40d	5	0.162	0.406	34
Style B				
2d	1	0.058	0.172	1,250
3d	1-⅛	0.062	0.188	980
4d	1-⅜	0.067	0.203	680
5d	1-⅝	0.072	0.219	510
6d	1-⅞	0.086	0.250	315
7d	2-⅛	0.086	0.250	280
8d	2-⅜	0.099	0.286	190
9d	2-⅝	0.099	0.286	170
10d	2-⅞	0.113	0.297	120

BRADS

Blunt Point   Diamond Point

Length (inch)	Body Diameter (inch)	Weight (number per pound)
3/8	0.035	9,520
1/2	0.035	7,080
1/2	0.048	3,990
5/8	0.035	5,680
5/8	0.048	3,200
3/4	0.035	4,800
3/4	0.048	2,620
3/4	0.062	1,550
7/8	0.035	4,220
7/8	0.048	2,220
7/8	0.062	1,280
1	0.054	1,500
1	0.062	1,120
1	0.072	904
1-1/4	0.054	1,210
1-1/4	0.062	940
1-1/4	0.080	560
1-1/2	0.054	1,040
1-1/2	0.080	470
1-1/2	0.099	320
1-3/4	0.062	670
1-3/4	0.080	400
1-3/4	0.099	270
2	0.062	580
2	0.080	350
2	0.113	180
2-1/4	0.080	320
2-1/4	0.113	160
2-1/2	0.080	290
2-1/2	0.131	110
2-3/4	0.131	97
3	0.148	70
3-1/4	0.148	65
3-1/2	0.162	50
4	0.192	31
4-1/2	0.207	24
5	0.225	18
5-1/2	0.244	14
6	0.262	11

SHEET METAL SCREW CHARACTERISTICS

Screw Size #	Screw Diameter (inches)	Diameter of Pierced Hole (inches)	Hole Size	Thickness of Metal– Gauge #
#4	.112	.086	#44	28
		.086	#44	26
		.093	#42	24
		.098	#42	22
		.100	#40	20
#6	.138	.111	#39	28
		.111	#39	26
		.111	#39	24
		.111	#38	22
		.111	#36	20
#7	.155	.121	#37	28
		.121	#37	26
		.121	#35	24
		.121	#33	22
		.121	#32	20
		–	#31	18
#8	.165	.137	#33	26
		.137	#33	24
		.137	#32	22
		.137	#31	20
		–	#30	18
#10	.191	.158	#30	26
		.158	#30	24
		.158	#30	22
		.158	#29	20
		.158	#25	18
#12	.218	–	#26	24
		.185	#25	22
		.185	#24	20
		.185	#22	18
#14	.251	–	#15	24
		.212	#12	22
		.212	#11	20
		.212	#9	18

STANDARD WOOD SCREW CHARACTERISTICS (Inches)

Screw Size #	Wood Screw Standard Lengths	Size of Pilot Hole		Size of Shank Hole	
		Softwood Bit #	Hardwood Bit #	Clearance Bit #	Hole Diameter
0	1/4"	75	66	52	.060"
1	1/4" to 3/8"	71	57	47	.073"
2	1/4" to 1/2"	65	54	42	.086"
3	1/4" to 5/8"	58	53	37	.099"
4	3/8" to 3/4"	55	51	32	.112"
5	3/8" to 3/4"	53	47	30	.125"
6	3/8" to 1 1/2"	52	44	27	.138"
7	3/8" to 1 1/2"	51	39	22	.151"
8	1/2" to 2"	48	35	18	.164"
9	5/8" to 2 1/4"	45	33	14	.177"
10	5/8" to 2 1/2"	43	31	10	.190"
11	3/4" to 3"	40	29	4	.203"
12	7/8" to 3 1/2"	38	25	2	.216"
14	1" to 4 1/2"	32	14	D	.242"
16	1 1/4" to 5 1/2"	29	10	I	.268"
18	1 1/2" to 6"	26	6	N	.294"
20	1 3/4" to 6"	19	3	P	.320"
24	3 1/2" to 6"	15	D	V	.372"

ALLEN HEAD AND MACHINE SCREW BOLT AND TORQUE CHARACTERISTICS

Number of Threads Per Inch	Allen Head and Mach. Screw Bolt Size	Allen Head Case H Steel 160,000 psi	Mach. Screw Yellow Brass 60,000 psi	Mach. Screw Silicone Bronze 70,000 psi
		Torque in Foot-Pounds or Inch-Pounds		
4.5	2"	8,800	–	–
5	1¾"	6,100	–	–
6	1½"	3,450	655	595
6	1⅜"	2,850	–	–
7	1¼"	2,130	450	400
7	1⅛"	1,520	365	325
8	1"	970	250	215
9	⅞"	640	180	160
10	¾"	400	117	104
11	⅝"	250	88	78
12	9/16"	180	53	49
13	½"	125	41	37
14	7/16"	84	30	27
16	⅜"	54	20	17
18	5/16"	33	125 in#	110 in#
20	¼"	16	70 in#	65 in#
24	#10	60	22 in#	20 in#
32	#8	46	19 in#	16 in#
32	#6	21	10 in#	8 in#
40	#5	–	7.2 in#	6.4 in#
40	#4	–	4.9 in#	4.4 in#
48	#3	–	3.7 in#	3.3 in#
56	#2	–	2.3 in#	2 in#

For fine thread bolts, increase by 9%.

HEX HEAD BOLT AND TORQUE CHARACTERISTICS

Bolt make-up is steel with coarse threads.

Number of Threads Per Inch	Hex Head Bolt Size	SAE 0-1-2 74,000 psi	SAE Grade 3 100,000 psi	SAE Grade 5 120,000 psi
		Torque = Foot-Pounds		
4.5	2"	2,750	5,427	4,550
5	1¾"	1,900	3,436	3,150
6	1½"	1,100	1,943	1,775
6	1⅜"	900	1,624	1,500
7	1¼"	675	1,211	1,105
7	1⅛"	480	872	794
8	1"	310	551	587
9	⅞"	206	372	382
10	¾"	155	234	257
11	⅝"	96	145	154
12	9/16"	69	103	114
13	½"	47	69	78
14	7/16"	32	47	54
16	⅜"	20	30	33
18	5/16"	12	17	19
20	¼"	6	9	10

For fine thread bolts, increase by 9%.

ANCHOR STRENGTH IN CONCRETE, HOLLOW CONCRETE BLOCK, AND BRICK

Pin Drive or Drive Nail Expansion Anchors, ZINC Body and Steel Nail

Anchor Diameter (inch)	Drill Hole Diameter (inch)	Allowable Tension Loads in Pounds In:			
		Anchor Depth	Concrete	Hollow Block	Solid Brick
3/16	3/16	1/2	100	—	—
		5/8	110	225	—
		3/4	115	68	115
1/4	1/4	1	225	—	—
		1-1/4	250	—	—
		1-3/4	300	—	—
		1-1/2	—	241	254

All figures approximate.

ANCHOR STRENGTH IN CONCRETE, HOLLOW CONCRETE BLOCK, AND BRICK *(cont.)*

Pin Drive or Drive Nail Expansion Anchors, NYLON Body and Steel Nail

Anchor Diameter (inch)	Hole Diameter (inch)	Allowable Tension Loads in Pounds In:			
		Anchor Depth (inch)	Concrete	Hollow Block	Solid Brick
3/16	3/16	3/4	50	43	39
		1	55	45	42
		1/2	60		
1/4	1/4	3/4	55	40	50
		1	58	43	55
		1-1/4	60	45	60
		1-1/2	60	50	63
		2	64	53	67

All figures approximate.

ANCHOR STRENGTH IN CONCRETE, HOLLOW CONCRETE BLOCK, AND BRICK *(cont.)*

Concrete Screws, Steel

Anchor Diameter (inch)	Drill Hole Diameter (inch)	Allowable Tension Loads in Pounds In:			
		Anchor Depth (inch)	Concrete	Hollow Block	Solid Brick
3/16	5/32	1	165	160	155
		1-1/4	218	190	268
		1-1/2	260	248	305
		1-3/4	330	303	333
1/4	3/16	1	220	95	90
		1-1/4	265	160	150
		1-1/2	440	179	162
		1-3/4	540	296	280
1/4	3/16	1-1/4	273	220	290
		1-1/2	380	273	330
		1-3/4	488	323	355

All figures approximate.

ANCHOR STRENGTH IN CONCRETE, HOLLOW CONCRETE BLOCK, AND BRICK *(cont.)*

Insert Anchors for Screws, Plastic

Anchor Size (inch)	Drill Hole Diameter (inch)	Allowable Tension Loads in Pounds In:			
		Anchor Depth (inch)	Concrete	Hollow Block	Solid Brick
#6 - #8 × 3/4	3/16	3/4	46	45	25
#8 - #10 × 7/8	3/16	7/8	75	73	40
#8 × 1	3/16				
#10 × 1	3/16				
#10 - #12 × 1	1/4	1	88	88	70
#14 × 1-1/2	5/16				
#14 - #16 × 1-1/2	5/16	1-1/2	210	210	220

All figures approximate.

ANCHOR STRENGTH IN CONCRETE, HOLLOW CONCRETE BLOCK, AND BRICK *(cont.)*

Double-Acting Expansion SHIELD

Anchor Size (inch)	Bolt Size (inch)	Drill Hole Diameter (inch)	Depth (inch)	Allowable Tension Loads in Pound For Anchors Embedded in Concrete
1/4	1/4	1/2	1-3/8	458
			1-1/2	470
			1-5/8	480
5/16	5/16	5/8	1-5/8	543
			1-7/8	560
		3/4	2-1/8	650
3/8	3/8	3/4	2	810
			2-1/8	920
1/2	1/2	7/8	2-1/2	1,643
			2-5/8	1,700
5/8	5/8	1	2-3/4	2,000
			3	2,150
3/4	3/4	1-1/4	3-1/2	3,000
			4	3,000

All values approximate.

ANCHOR STRENGTH IN CONCRETE, HOLLOW CONCRETE BLOCK, AND BRICK *(cont.)*

Anchor Strength For Lag Screws and Shields

Screw Diameter (inch)	Screw Diameter (inch)	Drill Hole Diameter (inch)	Depth (inch)	Allowable Tension Loads in Concrete
1/4	1/4	1/2	1	—
				94
				—
5/16	5/16	1/2	1-1/4	120
				—
			1-1/2	150
3/8	3/8	5/8	1-3/4	—
				245
				—
1/2	1/2	3/4	2	—
				290
				—
5/8	5/8	7/8	2	365
				—
3/4	3/4	1	2	545
				—

All values approximate.

ANCHOR STRENGTH IN CONCRETE, HOLLOW CONCRETE BLOCK, AND BRICK *(cont.)*

Long Lag Shield Anchor

Anchor Size (inch)	Bolt Size (inch)	Drill Hole Diameter (inch)	Embedment Depth (inch)	Allowable Tension Loads in Concrete
1/4	1/4	1/2	1-1/2	—
				120
				—
5/16	5/16	1/2	1-3/4	—
				235
				—
3/8	3/8	5/8	2-1/2	—
				295
				—
1/2	1/2	3/4	3	—
				580
				—
5/8	5/8	7/8	3-1/2	621
				—
3/4	3/4	1	3-1/2	815

All values approximate.

ROPE WEIGHTS AND STRENGTHS

Manila Rope, 3-Strand

Diameter		Strength		Weight	
Inch	mm	Minimum Breaking Strength Pounds	Safe Load Pounds	lb/ft	kg/m
3/16	5	405	35	0.014	0.020
1/4	6	540	45	0.018	0.027
5/16	8	900	75	0.026	0.039
3/8	10	1,215	101	0.038	0.056
7/16	11	1,575	131	0.049	0.072
1/2	12	2,385	199	0.070	0.104
9/16	14	3,105	259	0.096	0.143
5/8	16	3,960	330	0.127	0.189
3/4	18	4,860	405	0.159	0.237
13/16	20	5,850	488	0.186	0.277
7/8	22	6,930	578	0.214	0.318
1	24	8,100	675	0.257	0.382
1-1/16	26	9,450	788	0.298	0.443
1-1/8	28	10,800	900	0.343	0.510
1-1/4	30	12,150	1,010	0.397	0.591
1-3/8	32	13,500	1,130	0.456	0.679
1-1/2	36	16,650	1,390	0.570	0.848
1-5/8	40	20,250	1,690	0.711	1.06
1-3/4	44	23,850	1,990	0.850	1.26
2	48	27,900	2,330	1.02	1.52
2-1/8	52	32,400	2,700	1.20	1.79
2-1/4	56	36,900	3,080	1.39	2.07
2-1/2	60	42,300	3,530	1.64	2.44
2-5/8	64	46,800	3,900	1.82	2.71
2-7/8	68	54,900	4,580	2.15	3.20
3	72	57,500	4,790	2.30	3.42
3-5/16	80	69,500	5,790	2.84	4.23
3-5/8	88	81,900	6,830	3.49	5.19
4	96	94,500	7,880	4.14	6.16

ROPE WEIGHTS AND STRENGTHS *(cont.)*

Nylon Rope, 3-Strand and 8-Strand

Diameter		Strength		Weight	
Inch	mm	Minimum Breaking Strength Pounds	Safe Load Pounds	lb/ft	kg/m
3/16	5	880	73.3	0.009	0.013
1/4	6	1,486	124	0.016	0.023
5/16	8	2,295	191	0.025	0.036
3/8	10	3,240	270	0.036	0.053
7/16	11	4,320	360	0.048	0.071
1/2	12	5,670	473	0.063	0.094
9/16	14	7,200	600	0.080	0.119
5/8	16	8,910	743	0.099	0.147
3/4	18	12,780	1,070	0.143	0.213
7/8	22	17,280	1,440	0.195	0.290
1	24	22,230	1,850	0.253	0.377
1-1/16	26	25,200	2,100	0.287	0.427
1-1/8	28	28,260	2,360	0.322	0.479
1-1/4	30	34,830	2,900	0.397	0.591
1-5/16	32	38,250	3,190	0.437	0.650
1-1/2	36	48,600	4,050	0.570	0.848
1-5/8	40	57,375	4,780	0.673	1.00
1-3/4	44	66,150	5,510	0.780	1.16
2	48	84,600	7,050	1.00	1.49
2-1/8	52	95,400	7,950	1.13	1.68
2-1/4	56	107,100	8,930	1.27	1.89
2-1/2	60	131,400	11,000	1.57	2.34
2-5/8	64	144,000	12,000	1.73	2.57
2-3/4	68	171,000	14,300	2.08	3.10
3	72	185,400	15,500	2.26	3.36
3-1/4	80	224,100	18,700	2.75	4.09
3-1/2	88	267,300	22,300	3.29	4.90
4	96	324,000	27,000	4.00	5.95
4-1/4	104	369,000	30,800	4.60	6.85

ROPE WEIGHTS AND STRENGTHS *(cont.)*

Nylon Rope, 3-Strand and 8-Strand *(cont.)*

Diameter		Strength		Weight	
		Minimum Breaking Strength	Safe Load		
Inch	mm	Pounds	Pounds	lb/ft	kg/m
4-½	112	418,500	34,900	5.25	7.81
5	120	480,600	40,100	6.10	9.08
5-5/16	128	532,800	44,400	6.85	10.2
5-⅝	136	589,500	49,100	7.67	11.4
6	144	660,600	55,100	8.70	12.9

Polyester Rope, 3-Strand and 8-Strand

Rope Diameter		Strength		Weight	
		Minimum Breaking Strength	Safe Load		
Inch	mm	Pounds	Pounds	lb/ft	kg/m
3/16	5	765	63.8	0.011	0.016
¼	6	1,315	110	0.020	0.029
5/16	8	2,050	171	0.031	0.045
⅜	10	2,900	242	0.044	0.065
7/16	11	3,915	326	0.059	0.088
½	12	5,085	424	0.077	0.115
9/16	14	6,435	536	0.098	0.146
⅝	16	7,825	652	0.120	0.179
¾	18	11,200	933	0.172	0.256
⅞	22	15,225	1,270	0.234	0.348
1	24	19,775	1,650	0.304	0.452
1-1/16	26	22,225	1,850	0.342	0.509
1-⅛	28	24,800	2,070	0.385	0.578
1-¼	30	29,800	2,480	0.465	0.692
1-5/16	32	32,500	2,710	0.51	0.759
1-½	36	42,200	3,520	0.67	0.997
1-⅝	40	49,250	4,100	0.78	1.16
1-¾	44	57,000	4,750	0.91	1.35

ROPE WEIGHTS AND STRENGTHS *(cont.)*

Polyester Rope, 3-Strand and 8-Strand *(cont.)*

Rope Diameter		Strength		Weight	
		Minimum Breaking Strength	Safe Load		
Inch	mm	Pounds	Pounds	lb/ft	kg/m
2	48	72,000	6,000	1.17	1.74
2-⅛	52	81,000	6,750	1.33	1.98
2-¼	56	90,500	7,540	1.49	2.22
2-½	60	110,000	9,170	1.84	2.74
2-⅝	64	121,000	10,100	2.03	3.02
2-¾	68	144,000	12,000	2.43	3.62
3	72	156,000	13,000	2.64	3.93
3-¼	80	188,500	15,700	3.23	4.81
3-½	88	225,000	18,800	3.87	5.76
4	96	270,000	22,500	4.70	6.99
4-¼	104	310,000	25,800	5.47	8.14
4-½	112	355,000	29,600	6.30	9.38
5	120	410,000	34,200	7.32	10.9
5-5/16	128	459,000	38,300	8.25	12.3
5-⅝	136	508,500	42,400	9.25	13.8
6	144	567,000	47,300	10.5	15.6

Polyester/Polyolefin Dual Fiber Rope, 3-Strand

Rope Diameter		Strength		Weight	
		Minimum Breaking Strength	Safe Load		
Inch	mm	Pounds	Pounds	lb/ft	kg/m
¼	6	1,200	100	0.016	0.024
5/16	8	1,870	156	0.025	0.037
⅜	10	2,700	225	0.036	0.054
7/16	11	3,500	292	0.048	0.071
½	12	4,400	367	0.062	0.092
9/16	14	5,200	433	0.079	0.118
⅝	16	6,100	508	0.095	0.141

ROPE WEIGHTS AND STRENGTHS *(cont.)*

Polyester/Polyolefin Dual Fiber Rope, 3-Strand *(cont.)*

Rope Diameter		Strength		Weight	
Inch	mm	Minimum Breaking Strength Pounds	Safe Load Pounds	lb/ft	kg/m
¾	18	8,400	700	0.135	0.201
⅞	22	11,125	927	0.180	0.268
1	24	13,175	1,100	0.218	0.324
1-1/16	26	14,775	1,230	0.245	0.365
1-⅛	28	16,325	1,360	0.271	0.403
1-¼	30	19,900	1,660	0.334	0.497
1-5/16	32	21,950	1,830	0.365	0.543
1-½	36	28,250	2,350	0.470	0.699
1-⅝	40	32,950	2,750	0.550	0.818
1-¾	44	36,850	3,070	0.620	0.923
2	48	48,050	4,000	0.810	1.21
2-⅛	52	53,950	4,500	0.910	1.35
2-¼	56	59,950	5,000	1.01	1.50
2-½	60	73,550	6,130	1.24	1.85
2-⅝	64	80,650	6,720	1.36	2.02
2-¾	68	95,400	7,950	1.61	2.40
3	72	102,900	8,580	1.74	2.59
3-¼	80	122,800	10,200	2.12	3.15
3-½	88	144,800	12,100	2.50	3.72
4	96	171,000	14,300	3.00	4.46
4-¼	104	195,800	16,300	3.45	5.13
4-½	112	224,800	18,700	3.95	5.88
5	120	254,700	21,200	4.55	6.77
5-5/16	128	282,600	23,600	5.06	7.53
5-⅝	136	312,300	26,000	5.62	8.36
6	144	351,000	29,300	6.35	9.45

ROPE WEIGHTS AND STRENGTHS *(cont.)*

Polypropylene Fiber Rope, 3-Strand and 8-Strand

Rope Diameter		Strength		Weight	
Inch	mm	Minimum Breaking Strength Pounds	Safe Load Pounds	lb/ft	kg/m
3/16	5	650	54.2	0.01	0.01
¼	6	1,125	93,8	0.01	0.02
5/16	8	1,710	143	0.02	0.03
⅜	10	2,430	203	0.03	0.04
7/16	11	3,150	263	0.04	0.05
½	12	3,780	315	0.05	0.07
9/16	14	4,590	383	0.06	0.09
⅝	16	5,580	465	0.07	0.11
¾	18	7,650	638	0.10	0.15
⅞	22	10,350	863	0.14	0.21
1	24	12,825	1,070	0.18	0.27
1-1/16	26	14,400	1,200	0.20	0.30
1-⅛	28	16,000	1,330	0.23	0.34
1-¼	30	19,350	1,610	0.28	0.41
1-5/16	32	21,150	1,760	0.30	0.45
1-½	36	27,350	2,280	0.39	0.59
1-⅝	40	31,950	2,660	0.46	0.68
1-¾	44	36,900	3,080	0.53	0.79
2	48	46,800	3,900	0.69	1.03
2-⅛	52	52,650	4,390	0.78	1.16
2-¼	56	59,400	4,950	0.88	1.31
2-½	60	72,000	6,000	1.07	1.59
2-⅝	64	80,500	6,710	1.20	1.79
2-¾	68	94,500	7,880	1.41	2.10
3	72	102,600	8,550	1.53	2.28
3-¼	80	121,500	10,100	1.86	2.77
3-½	88	144,000	12,000	2.23	3.32
4	96	171,900	14,300	2.72	4.05
4-¼	104	198,000	16,500	3.15	4.69
4-½	112	223,200	18,600	3.60	5.36
5	120	256,500	21,400	4.20	6.25
5-5/16	128	287,100	23,900	4.74	7.05
5-⅝	136	319,500	26,600	5.31	7.90
6	144	358,200	29,900	6.03	8.97

ROPE WEIGHTS AND STRENGTHS *(cont.)*

Sisal Rope, 3-Strand

Nominal Rope Diameter		Strength		Weight	
Inch	mm	Minimum Breaking Strength Pounds	Safe Load Pounds	lb/ft	kg/m
3/16	5	290	24.2	0.01	0.02
1/4	6	385	32.1	0.02	0.03
5/16	8	640	53.3	0.03	0.04
3/8	10	865	72.1	0.04	0.06
7/16	11	1,120	93.3	0.05	0.07
1/2	12	1,700	142	0.07	0.10
9/16	14	2,210	184	0.10	0.14

CHAIN SAW CLASSIFICATIONS

1. Mini–saw: Weight, 6 to 9 pounds. Engine size, 1.8 to 2.5 cubic inches (or electric). Bar lengths, 8 to 12 inches, 1/4 inch pitch.
2. Light–Duty: Weight, 9 to 13 pounds. Engine size, 2.5 to 3.8 cubic inches (or electric), and bar lengths 14 to 16 inches, 3/8 inch pitch.
3. Medium–Duty: Weight, 13 to 18 pounds. Engine size, 3.5 to 4.8 cubic inches. Bar lengths, 16 to 24 inches, 3/8 to 0.404 inch pitch.
4. Heavy–Duty: Weight over 18 pounds. Engine size over 4.8 cubic inches. Bar lengths over 24 inches, 0.404 to 1/2 inch pitch.

DEPTH CAPACITY OF CIRCULAR POWER SAWS

Blade Diameter	Capacity @ 90°	Capacity @ 45°
4-1/2	1-5/16	1-1/16 to 1-1/14
6-1/2	2-1/16	1-5/8
6-3/4	2-7/32	1-3/4
7-1/4	2-3/8 to 2-7/16	1-7/8 to 1-29/32
7-1/2	2-17/22	2-1/16
8-1/4	2-15/16	2-1/4
10-1/4	3-5/8	2-3/4
12	4-3/8	3-5/16

CIRCULAR SAW BLADES

Abrasive Wheel
Abrasive blade made of aluminum oxide (metal cutting) or silicon carbide (masonry cutting). Standard sizes of 6-, 7-, and 8-inch diameter and with ½- or ⅝-diameter arbor. No teeth.

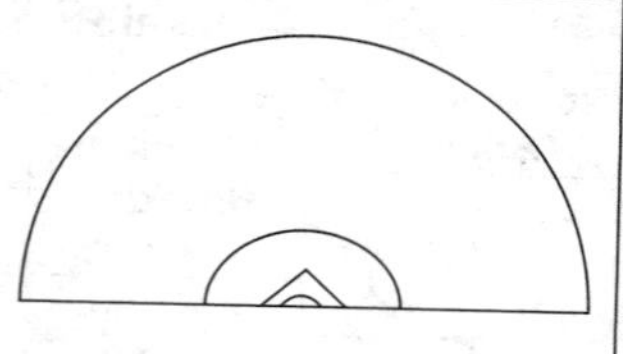

Combination Blade
Leaves a very smooth finish. For crosscuts and miters across wood grain.

Ripping Blade
Large, set teeth with deep gullets. For cutting fast in the direction of the wood grain. Very rough finish.

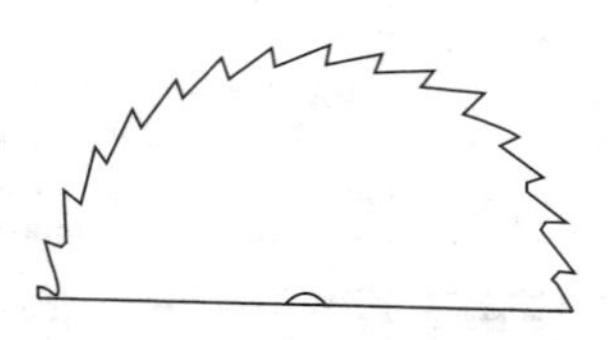

Chisel–Tooth Combination Blade
Good for both ripping and crosscuts. Cuts fast, but leaves a rough cut. Durable blades.

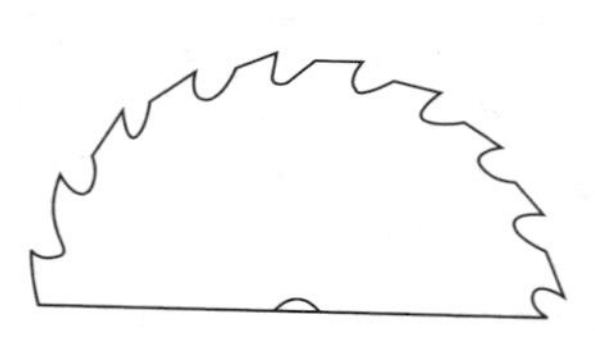

CIRCULAR SAW BLADES *(cont.)*

Crosscut, Fine–Tooth, and Paneling Blades

A large number of sharp teeth. Designed for cutting across the wood grain and leaving a smooth edge. Paneling blades have many, extra fine teeth.

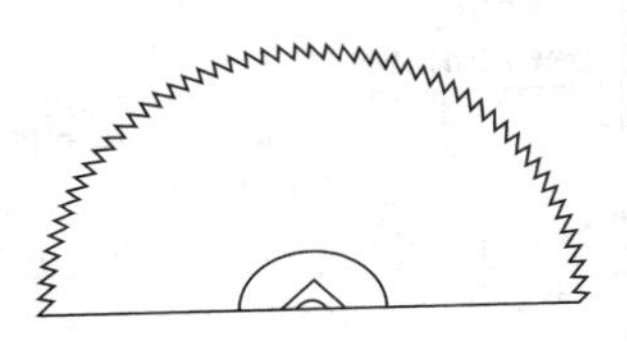

Flattop–Ground Carbide Tipped Blade

Fast cutting, long lasting blade. Good for ripping, crosscutting, and mitering but does not leave a smooth edge.

Steel Cutting Blade

Used to cut iron and steel up to 3⁄32-inch thick.

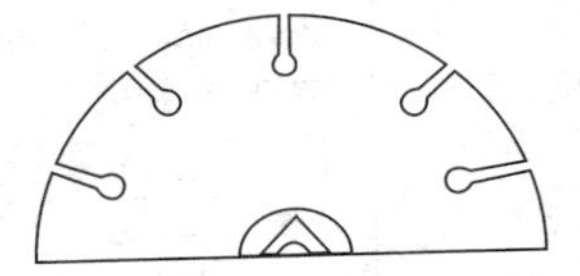

SABER SAW BLADES	
Teeth per inch	**Usage**
3	Lumber up to 6 inches thick, fast cutting, very rough cut.
5 or 6	Lumber up to 2 inches thick, fast cutting, rough cut.
7 or 8	Best general purpose blade; relatively smooth cut. Good for lumber and fiber insulation board.
10	Good general purpose blade. For hard, abrasive materials, such as laminates, use metal-cutting H.S.S.
12 or 14	Very smooth cutting but very slow. Good for hardwoods, plywood, fiberglass, plastics, rubber, linoleum, laminates, and plexiglass. If the material is particularly hard or abrasive, use metal cutting H.S.S. blades.
Knife	Either a knife edge or a sharp edge with an abrasive bonded to the blade. No-grit blades for cutting rubber, cork, leather, cardboard, styrofoam, and silicones. Grit blades for fiberglass, epoxies, ceramic tile, stone, clay pipe, brick, steel, and veneer.
Metal Cutting Blades	
6 to 10	For aluminum, brass, copper, laminates, hardwoods, and other soft materials up to ½-inch thickness.
14	Same as above, plus mild steels and hardboards. Leaves a much smoother edge. Thickness of ¼- to ½-inch maximum.
18	Same as 14 tpi but maximum thickness ⅛-inch.
24	Smooth edge cutting for steel and sheet metal. Also good for other hard materials with a maximum thickness of ⅛-inch.
32	Very fine cuts for steel and thin wall tubing up to 1/16-inch thick.

BAND SAW TEETH PER INCH AND SPEED

Type of Material To Be Cut	Size of Material			
	½"–1"	1"–2"	½"–1"	1"–2"
	Teeth-per-inch		Speed (fpm)	
Steels				
Angle Iron	14	14	190	175
Armor plate	14	12	100	75
Cast Iron	12	10	200	185
Cast steels	14	12	150	75
Graphic steel	14	12	150	125
High-speed steel	14	10	100	75
I-beams and channels	14	14	250	200
Pipe	14	12	250	225
Stainless steel	12	10	60	50
Tubing (thinwall)	14	14	250	200
Nonferrous Metals				
Aluminum (all types)	8	6	250	250
Beryllium	10	8	175	150
Brass	8	8	250	250
Bronze (cast)	10	8	185	125
Bronze (rolled)	12	10	175	125
Copper	10	8	250	225
Magnesium	8	8	250	250

RECOMMENDED DRILLING SPEEDS (RPMS)		
Material	**Bit Sizes**	**RPM Speed Range**
Glass	**Special Metal Tube Drilling**	**700**
Plastics	7/16" and larger 3/8" 5/16" 1/4" 3/16" 1/8" 1/16" and smaller	500–1,000 1,500–2,000 2,000–2,500 3,000–3,500 3,500–4,000 5,000–6,000 6,000–6,500
Woods	1" and larger 3/4" to 1" 1/2" to 3/4" 1/4" to 1/2" 1/4" and smaller carving / routing	700–2,000 2,000–2,300 2,300–3,100 3,100–3,800 3,800–4,000 4,000–6,000
Soft Metals	7/16" and larger 3/8" 5/16" 1/4" 3/16" 1/8" 1/16" and smaller	1,500–2,500 3,000–3,500 3,500–4,000 4,500–5,000 5,000–6,000 6,000–6,500 6,000–6,500
Steel	7/16" and larger 3/8" 5/16" 1/4" 3/16" 1/8" 1/16" and smaller	500–1,000 1,000–1,500 1,000–1,500 1,500–2,000 2,000–2,500 3,000–4,000 5,000–6,500
Cast Iron	7/16" and larger 3/8" 5/16" 1/4" 3/16" 1/8" 1/16" and smaller	1,000–1,500 1,500–2,000 1,500–2,000 2,000–2,500 2,500–3,000 3,500–4,500 6,000–6,500

TORQUE LUBRICATION EFFECTS IN FOOT-POUNDS

Lubricant	5/16" – 18 Thread	1/2" – 13 Thread	Torque Decrease
Graphite	13	62	49 – 55%
Mily Film	14	66	45 – 52%
White Grease	16	79	35 – 45%
SAE 30	16	79	35 – 45%
SAE 40	17	83	31 – 41%
SAE 20	18	87	28 – 38%
Plated	19	90	26 – 34%
No Lube	29	121	0%

METALWORKING LUBRICANTS

Materials	Threading	Lathing	Drilling
Machine Steels	Dissolvable Oil Mineral Oil Lard Oil	Dissolvable Oil	Dissolvable Oil Sulpherized Oil Min. Lard Oil
Tool Steels	Lard Oil Sulpherized Oil	Dissolvable Oil	Dissolvable Oil Sulpherized Oil
Cast Irons	Sulpherized Oil Dry Min. Lard Oil	Dissolvable Oil Dry	Dissolvable Oil Dry Air Jet
Malleable Irons	Soda Water Lard Oil	Soda Water Dissolvable Oil	Soda Water Dry
Aluminums	Kerosene Dissolvable Oil Lard Oil	Dissolvable Oil	Kerosene Dissolvable Oil
Brasses	Dissolvable Oil Lard Oil	Dissolvable Oil	Kerosene Dissolvable Oil Dry
Bronzes	Dissolvable Oil Lard Oil	Dissolvable Oil	Dissolvable Oil Dry
Coppers	Dissolvable Oil Lard Oil	Dissolvable Oil	Kerosene Dissolvable Oil Dry

STEEL SHEET GAUGES

Gauge Number	Steel Weight (lbs per sq foot)	Thickness Inches U.S. Standard Gauge	Gauge Number	Steel Weight (lbs per sq foot)	Thickness Inches U.S. Standard Gauge
7/0	20.00	0.5000	20	1.50	0.0375
6/0	18.75	0.4687	21	1.37	0.0344
5/0	17.50	0.4375	22	1.25	0.0312
4/0	16.25	0.4062	23	1.12	0.0281
3/0	15.00	0.3750	24	1.00	0.0250
2/0	13.75	0.3437	25	0.875	0.0219
0	12.50	0.3125	26	0.750	0.0187
1	11.25	0.2812	27	0.687	0.0172
2	10.62	0.2656	28	0.625	0.0156
3	10.00	0.2500	29	0.562	0.0141
4	9.37	0.2344	30	0.500	0.0125
5	8.75	0.2187	31	0.437	0.0109
6	8.12	0.2031	32	0.406	0.0102
7	7.50	0.1875	33	0.375	0.0094
8	6.87	0.1719	34	0.344	0.0086
9	6.25	0.1562	35	0.312	0.0078
10	5.62	0.1406	36	0.281	0.0070
11	5.00	0.1250	37	0.266	0.0066
12	4.37	0.1094	38	0.250	0.0062
13	3.75	0.0937	39	0.234	0.0059
14	3.12	0.0781	40	0.219	0.0055
15	2.81	0.0703	41	0.211	0.0053
16	2.50	0.0625	42	0.203	0.0051
17	2.25	0.0562	43	0.195	0.0049
18	2.00	0.0500	44	0.187	0.0047
19	1.75	0.0437			

STEEL PLATE SIZES

Thickness (inches)	Weight (lbs/sq foot)	Thickness (inches)	Weight (lbs/sq foot)
3/16	7.65	2-1/8	86.70
1/4	10.20	2-1/4	91.80
5/16	12.75	2-1/2	102.00
3/8	15.30	2-3/4	112.20
7/16	17.85	3	122.40
1/2	20.40	3-1/4	132.60
9/16	22.95	3-1/2	142.80
5/8	25.50	3-3/4	153.00
11/16	28.05	4	163.20
3/4	30.60	4-1/4	173.40
13/16	33.15	4-1/2	183.60
7/8	35.70	5	204.00
1	40.80	5-1/2	224.40
1-1/8	45.90	6	244.80
1-1/4	51.00	6-1/2	265.20
1-3/8	56.10	7	285.60
1-1/2	61.20	7-1/2	306.00
1-5/8	66.30	8	326.40
1-3/4	71.40	9	367.20
1-7/8	76.50	10	408.00
2	81.60		

SANDPAPER AND ABRASIVES

Grit	Description	Use
12 16 20 24 30 36	Very Coarse Very Coarse Very Coarse Very Coarse Coarse Coarse	Very rough work, usually requires high speed, heavy machines. For unplaned woods, wood floors, rough cut.
40 50 60	Coarse Coarse Medium	Rough wood work, #1 is coarsest for use with pad sander.
80 100 120	Medium Medium Fine	General wood work, plaster patches, first smooth of old paint.
150 180	Fine Fine	Hardwood prep, final for softwoods, old paint.
220 240 280	Very Fine Very Fine Very Fine	Final sanding or between coats, won't show sand marks, dry sanding.
320 360 400	Extra Fine Extra Fine Extra Fine	Polish final coats, between coats, wet sand paints and varnishes, top coats.
500 600	Super Fine Super Fine	Sand metal, plastic and ceramics, wet sanding.

GLUE CHARACTERISTICS

This guide is reliable but not infallible: Because of the great variation between the formulations of various manufacturers, one may work, another fail. The manufacturer's label should help you in making your selection.

Type	Uses	Holding Power (lbs/sq. in.)	Water Resistance	Set Time	Cure Time
Acrylic adhesive	Bonds most surfaces, including oily or porous. Nonflammable.	6,000	Good	30–60 seconds	45 minutes
Acrylonitrile	Carpet and fabrics, glass and metal. Good oil resistance.	2,000–3,000	Good	Varies	Variable
Aliphatic	Excellent for bonding wood. Quick tack. Some creep.	2,000–3,500	Low	30 minutes	24 hours
Anaerobic adhesive	Sealing nuts to bolts. Cannot be used on polypropylene. Protects against corrosion.	2,000–3,000	Excellent	Varies in absence of air	Variable
Casein glue	Heavy wood gluing. Moisture-resistant but mold-susceptible.	2,500	Good	3 hours at 70° or above	24 hours
Cellulose	Repairs on furniture, ceramics, glass, fabrics, and plastics.	3,500	Excellent	2 hours	24–48 hours

GLUE CHARACTERISTICS *(cont.)*

Type	Uses	Holding Power (lbs/sq. in.)	Water Resistance	Set Time	Cure Time
Clear cement	Model making.	2,000–3,000	Excellent	5 minutes to 1 hour	24 hours
Contact cement	Bonds veneer to cabinet or counter tops. Also bonds plastic, foam, hardboard, or metal to wood.	—	Variable	Dry surfaces separately. Sets on contact.	48 hours to several weeks
Epoxy	Bonds nonporous materials, metal, pipes, ceramic, china, marble, glass, and masonry. Not good on wood.	2,000–3,500	Excellent	4 minutes to 1 hour	4 minutes to 24 hours
Hide glue (liquid)	Conventional furniture-maker's glue. Sensitive to high humidity and moisture.	3,200	Low	1 hour	8 hours at 70°F
Instant-set glue (cyanoacrylates)	Nonporous material, ceramics, plastic, rubber, metal, synthetics. Will not bond to Teflon or polyethylene. Cannot be used on plastic foam.	3,900–8,000	Fair	30 seconds	12–48 hours

Latex mastic	Ceiling tile, floor tile, paneling.	50–500	Good	On contact to 2 hours	2–3 days
Plastic cement	Nonporous material, wood, fabric.	500	Good	5 minutes to 1 hour	6–24 hours
Plastic resin glue	Furniture repair, wood, hardboard, chipboard, paneling.	3,000	Excellent	3–5 hours	10–30 hours at 70°F or above
Polyvinyl acetate	All-purpose (except metal). Poor in heat or moisture.	3,200–4,000	Poor	20 minutes	8 hours at 70°F
Resorcinol	Hard and medium-hard wood, plywood.	135,000	Excellent	7–10 hours at 70°F or above	24–30 hours
Silicone rubber	All woods, metals, plastics, masonry, ceramic tile, and glass. Very durable.	4,500–9,000	Excellent	6–8 minutes	24–48 hours
Styrene-butadiene	Hard flooring, gypsum board, quarry tiles, and attaching fixtures to walls. Can be drilled, sanded, and painted. May not be freeze-resistant. Used for jointing PVC pipes.	—	Excellent	1–3 hours	48 hours

GLUE CHARACTERISTICS *(cont.)*

Type	Uses	Holding Power (lbs/sq. in.)	Water Resistance	Set Time	Cure Time
Urethane adhesive	Wood, plastic, metal, ceramics, and glass. Has strength of epoxy but takes longer to cure.	—	Good	1-3 hours	24 hours
Polyvinyl chloride	All-purpose.	3,200	Excellent	10–15 minutes	8–24 hours
Wood glue	Wood, paper, cork, plywood, and paneling.	—	Good	1–3 hours	18 hours
Wallpaper paste	Wallpaper.	—	Fair	NA	NA
Asphalt	Attaching acoustical, wall and flooring tile, rigid insulations, and vapor barriers.	—	Excellent	On contact	24 hours
Elastomerics	Paper, leather, vinyl, and rubber.	—	Excellent	On contact	—
Phenolic blends	Metal to metal, wood, tile, glass, and phenolic plastics.	—	Excellent	Varies	Varies

CHAPTER 11
Conversion Factors

COMMONLY USED CONVERSION FACTORS		
Multiply	**By**	**To Obtain**
Acres	43,560	Square feet
Acres	1.562×10^{-3}	Square miles
Acre-feet	43,560	Cubic feet
Amperes per sq. cm.	6.452	Amperes per sq. in.
Amperes per sq. in.	0.1550	Amperes per sq. cm.
Ampere-turns.	1.257	Gilberts
Ampere-turns per cm. . . .	2.540	Ampere-turns per in.
Ampere-turns per in.	0.3937	Ampere-turns per cm.
Atmospheres	76.0	Cm. of mercury
Atmospheres	29.92	Inches of mercury
Atmospheres	33.90	Feet of water
Atmospheres	14.70	Pounds per sq. in.
British thermal units.	252.0	Calories
British thermal units.	778.2	Foot-pounds
British thermal units.	3.960×10^{-4}	Horsepower-hours
British thermal units.	0.2520	Kilogram-calories
British thermal units.	107.6	Kilogram-meters
British thermal units.	2.931×10^{-4}	Kilowatt-hours
British thermal units.	1,055	Watt-seconds
B.t.u. per hour	2.931×10^{-4}	Kilowatts
B.t.u. per minute	2.359×10^{-2}	Horsepower
B.t.u. per minute	1.759×10^{-2}	Kilowatts
Bushels	1.244	Cubic feet
Centimeters	0.3937	Inches
Circular mils.	5.067×10^{-6}	Square centimeters
Circular mils.	0.7854×10^{-6}	Square inches
Circular mils.	0.7854	Square mils
Cords	128	Cubic feet
Cubic centimeters	6.102×10^{-6}	Cubic inches

COMMONLY USED CONVERSION FACTORS *(cont.)*

Multiply	By	To Obtain
Cubic feet	0.02832	Cubic meters
Cubic feet	7.481	Gallons
Cubic feet	28.32	Liters
Cubic inches	16.39	Cubic centimeters
Cubic meters	35.31	Cubic feet
Cubic meters	1.308	Cubic yards
Cubic yards	0.7646	Cubic meters
Degrees (angle)	0.01745	Radians
Dynes	2.248×10^{-6}	Pounds
Ergs	1	Dyne-centimeters
Ergs	7.37×10^{-6}	Foot-pounds
Ergs	10^{-7}	Joules
Farads	10^{6}	Microfarads
Fathoms	6	Feet
Feet	30.48	Centimeters
Feet of water	.08826	Inches of mercury
Feet of water	304.8	Kg. per square meter
Feet of water	62.43	Pounds per square ft.
Feet of water	0.4335	Pounds per square in.
Foot-pounds	1.285×10^{-2}	British thermal units
Foot-pounds	5.050×10^{-7}	Horsepower-hours
Foot-pounds	1.356	Joules
Foot-pounds	0.1383	Kilogram-meters
Foot-pounds	3.766×10^{-7}	Kilowatt-hours
Gallons	0.1337	Cubic feet
Gallons	231	Cubic inches
Gallons	3.785×10^{-3}	Cubic meters
Gallons	3.785	Liters
Gallons per minute	2.228×10^{-3}	Cubic feet per sec.
Gausses	6.452	Lines per square in.
Gilberts	0.7958	Ampere-turns
Henries	10^{3}	Millihenries
Horsepower	42.41	B.t.u. per min.
Horsepower	2,544	B.t.u. per hour

COMMONLY USED CONVERSION FACTORS *(cont.)*

Multiply	By	To Obtain
Horsepower	550	Foot-pounds per sec.
Horsepower	33,000	Foot-pounds per min.
Horsepower	1.014	Horsepower (metric)
Horsepower	10.70	Kg. calories per min.
Horsepower	0.7457	Kilowatts
Horsepower (boiler)	33,520	B.t.u. per hour
Horsepower-hours	2,544	British thermal units
Horsepower-hours	1.98×10^{6}	Foot-pounds
Horsepower-hours	2.737×10^{5}	Kilogram-meters
Horsepower-hours	0.7457	Kilowatt-hours
Inches	2.540	Centimeters
Inches of mercury	1.133	Feet of water
Inches of mercury	70.73	Pounds per square ft.
Inches of mercury	0.4912	Pounds per square in.
Inches of water	25.40	Kg. per square meter
Inches of water	0.5781	Ounces per square in.
Inches of water	5.204	Pounds per square ft.
Joules	9.478×10^{-4}	British thermal units
Joules	0.2388	Calories
Joules	10^{7}	Ergs
Joules	0.7376	Foot-pounds
Joules	2.778×10^{-7}	Kilowatt-hours
Joules	0.1020	Kilogram-meters
Joules	1	Watt-seconds
Kilograms	2.205	Pounds
Kilogram-calories	3.968	British thermal units
Kilogram meters	7.233	Foot-pounds
Kg per square meter	3.281×10^{-3}	Feet of water
Kg per square meter	0.2048	Pounds per square ft.
Kg per square meter	1.422×10^{-3}	Pounds per square in.
Kilolines	10^{3}	Maxwells
Kilometers	3.281	Feet
Kilometers	0.6214	Miles
Kilowatts	56.87	B.t.u. per min.

COMMONLY USED CONVERSION FACTORS *(cont.)*

Multiply	By	To Obtain
Kilowatts	737.6	Foot-pounds per sec.
Kilowatts	1.341	Horsepower
Kilowatts-hours	3409.5	British thermal units
Kilowatts-hours	2.655×10^6	Foot-pounds
Knots	1.152	Miles
Liters	0.03531	Cubic feet
Liters	61.02	Cubic inches
Liters	0.2642	Gallons
Log N_e or in N	0.4343	Log_{10} N
Log N	2.303	Log_e N or in N
Lumens per square ft.	1	Footcandles
Maxwells	10^{-3}	Kilolines
Megalines	10^6	Maxwells
Megaohms	10^6	Ohms
Meters	3.281	Feet
Meters	39.37	Inches
Meter-kilograms	7.233	Pound-feet
Microfarads	10^{-6}	Farads
Microhms	10^{-6}	Ohms
Microhms per cm. cube	0.3937	Microhms per in. cube
Microhms per cm. cube	6.015	Ohms per mil. foot
Miles	5,280	Feet
Miles	1.609	Kilometers
Miner's inches	1.5	Cubic feet per min.
Ohms	10^{-6}	Megohms
Ohms	10^6	Microhms
Ohms per mil foot	0.1662	Microhms per cm. cube
Ohms per mil foot	0.06524	Microhms per in. cube
Poundals	0.03108	Pounds
Pounds	32.17	Poundals
Pound-feet	0.1383	Meter-Kilograms
Pounds of water	0.01602	Cubic feet
Pounds of water	0.1198	Gallons
Pounds per cubic foot	16.02	Kg. per cubic meter
Pounds per cubic foot	5.787×10^{-4}	Pounds per cubic in.

COMMONLY USED CONVERSION FACTORS *(cont.)*

Multiply	By	To Obtain
Pounds per cubic inch. . .	27.68	Grams per cubic cm.
Pounds per cubic inch. . .	2.768×10^{-4}	Kg. per cubic meter
Pounds per cubic inch. . .	1.728	Pounds per cubic ft.
Pounds per square foot . .	0.01602	Feet of water
Pounds per square foot . .	4.882	Kg. per square meter
Pounds per square foot . .	6.944×10^{-3}	Pounds per sq. in.
Pounds per square inch . .	2.307	Feet of water
Pounds per square inch . .	2.036	Inches of mercury
Pounds per square inch . .	703.1	Kg. per square meter
Radians	57.30	Degrees
Square centimeters	1.973×10^{5}	Circular mils
Square feet	2.296×10^{-5}	Acres
Square feet	0.09290	Square meters
Square inches	1.273×10^{6}	Circular mils
Square inches	6.452	Square centimeters
Square kilometers	0.3861	Square miles
Square meters	10.76	Square feet
Square miles	640	Acres
Square miles	2.590	Square kilometers
Square millimeters	1.973×10^{3}	Circular mils
Square mils	1.273	Circular mils
Tons (long)	2,240	Pounds
Tons (metric)	2,205	Pounds
Tons (short)	2,000	Pounds
Watts	0.05686	B.t.u. per minute
Watts	10^{7}	Ergs per sec.
Watts	44.26	Foot-pounds per min.
Watts	1.341×10^{-3}	Horsepower
Watts	14.34	Calories per min.
Watts-hours	3.412	British thermal units
Watts-hours	2,655	Footpounds
Watts-hours	1.341×10^{-3}	Horsepower-hours
Watts-hours	0.8605	Kilogram-calories
Watts-hours	376.1	Kilogram-meters
Webers	10^{8}	Maxwells

DECIMAL EQUIVALENTS OF FRACTIONS

8ths	32nds	64ths	64ths
1/8 = .125	1/32 = .03125	1/64 = .015625	33/64 = .515625
1/4 = .250	3/32 = .09375	3/64 = .046875	35/64 = .546875
3/8 = .375	5/32 = .15625	5/64 = .078125	37/64 = .57812
1/2 = .500	7/32 = .21875	7/64 = .109375	39/64 = .609375
5/8 = .625	9/32 = .28125	9/64 = .140625	41/64 = .640625
3/4 = .750	11/32 = .34375	11/64 = .171875	43/64 = .671875
7/8 = .875	13/32 = .40625	13/64 = .203128	45/64 = .703125
16ths	15/32 = .46875	15/64 = .234375	47/64 = .734375
1/16 = .0625	17/32 = .53125	17/64 = .265625	49/64 = .765625
3/16 = .1875	19/32 = .59375	19/64 = .296875	51/64 = .796875
5/16 = .3125	21/32 = .65625	21/64 = .328125	53/64 = .828125
7/16 = .4375	23/32 = .71875	23/64 = .359375	55/64 = .859375
9/16 = .5625	25/32 = .78125	25/64 = .390625	57/64 = .890625
11/16 = .6875	27/32 = .84375	27/64 = .421875	59/64 = .921875
13/16 = .8125	29/32 = .90625	29/64 = .453125	61/64 = .953125
15/16 = .9375	31/32 = .96875	31/64 = .484375	63/64 = .984375

FRACTIONS OF INCHES—MILLIMETER EQUIVALENTS

Fraction of Inch	Millimeter Equivalent	Fraction of Inch	Millimeter Equivalent	Fraction of Inch	Millimeter Equivalent
1/64	0.39688	17/32	13.494	1-3/32	27.781
1/32	0.79375	35/64	13.891	1-1/8	28.575
3/64	1.1906	9/16	14.288	1-5/32	29.369
1/16	1.5875	37/64	14.684	1-3/16	30.163
5/64	1.9844	19/32	15.081	1-7/32	30.956
3/32	2.3813	39/64	15.478	1-1/4	31.750
7/64	2.7781	5/8	15.875	1-9/32	32.544
1/8	3.1750	41/64	16.272	1-5/16	33.338
9/64	3.5719	21/32	16.669	1-11/32	34.131
5/32	3.9688	43/64	17.066	1-3/8	34.925
11/64	4.3656	11/16	17.463	1-13/32	35.719
3/16	4.7625	45/64	17.859	1-7/16	36.513
13/64	5.1594	23/32	18.256	1-15/32	37.306
7/32	5.5563	47/64	18.653	1-1/2	38.100
15/64	5.9531	3/4	19.050	1-17/32	38.894
1/4	6.3500	49/64	19.447	1-9/16	39.688
17/64	6.7469	25/32	19.844	1-19/32	40.481
9/32	7.1438	51/64	20.241	1-5/8	41.275
19/64	7.5406	13/16	20.638	1-21/32	42.069
5/16	7.9375	53/64	21.034	1-11/16	42.863
21/64	8.3344	27/32	21.431	1-23/32	43.656
11/32	8.7313	55/64	21.828	1-3/4	44.450
23/64	9.1281	7/8	22.225	1-25/32	45.244
3/8	9.5250	57/64	22.622	1-13/16	46.038
25/64	9.9219	29/32	23.019	1-27/32	46.831
13/32	10.319	59/64	23.416	1-7/8	47.625
27/64	10.716	15/16	23.813	1-29/32	48.419
7/16	11.113	61/64	24.209	1-15/16	49.213
29/64	11.509	31/32	24.606	1-31/32	50.006
15/32	11.906	63/64	25.003	2	50.800
31/64	12.303	1	25.400		
1/2	12.700	1-1/32	26.194		
33/64	13.097	1-1/16	26.988		

COMMONLY USED GEOMETRICAL RELATIONSHIPS
Diameter of a circle × 3.1416 = Circumference. Radius of a circle × 6.283185 = Circumference. Square of the radius of a circle × 3.1416 = Area. Square of the diameter of a circle × 0.7854 = Area. Square of the circumference of a circle × 0.07958 = Area.
Half the circumference of a circle × half its diameter = Area. Circumference of a circle × 0.159155 = Radius. Square root of the area of a circle × 0.56419 = Radius. Circumference of a circle × 0.31831 = Diameter. Square root of the area of a circle × 1.12838 = Diameter.
Diameter of a circle × 0.866 = Side of an inscribed equilateral triangle. Diameter of a circle × 0.7071 = Side of an inscribed square. Circumference of a circle × 0.225 = Side of an inscribed square. Circumference of a circle × 0.282 = Side of an equal square. Diameter of a circle × 0.8862 = Side of an equal square.
Base of a triangle × one-half the altitude = Area. Multiplying both diameters and .7854 together = Area of an ellipse. Surface of a sphere × one-sixth of its diameter = Volume. Circumference of a sphere × its diameter = Surface. Square of the diameter of a sphere × 3.1416 = Surface.
Square of the circumference of a sphere × 0.3183 = Surface. Cube of the diameter of a sphere × 0.5236 = Volume. Cube of the circumference of a sphere × 0.016887= Volume. Radius of a sphere × 1.1547 = Side of an inscribed cube. Diameter of a sphere divided by √3 = Side of an inscribed cube.
Area of its base × one-third of its altitude = Volume of a cone or pyramid whether round, square, or triangular. Area of one of its sides × 6 = Surface of the cube. Altitude of trapezoid × one-half the sum of its parallel sides = Area.

CHAPTER 12
Glossary

GLOSSARY

A

ABS: Plastic pipe used for plumbing construction.

Abut: Joining end-to-end.

Accelerator: A concrete additive used to speed the curing time of concrete.

Acoustical: Referring to the study of sound transmission or reduction.

Adhesive: A bonding material.

Adjacent: Touching; next to.

Aggregate: Fine, lightweight, coarse, or heavyweight grades of sand, vermiculite, perlite, or gravel added to cement for concrete or plaster.

Air-drying: Method of removing excess moisture from lumber using natural circulation of air.

Air Handling Unit: A mechanical unit used for air conditioning or movement of air.

Allowable Load: Maximum supportable load of any construction component(s).

Allowable Span: Maximum length permissible for any framing component without support.

Anchor Bolt: A J- or L-shaped steel rod threaded on one end for securing structural members to concrete, etc.

Apron: A piece of window trim located beneath the window sill; also used to designate the front of a building.

Arbor: An axle on which a cutting tool is mounted; used in reference to the mounting of a circular saw blade.

Architect's Scale: A rule with scales indicating feet, inches, and fractions of inches.

Asphalt: The general term for a black material produced as a by-product of oil (asphalt) or coal (pitch or coal tar).

Asphalt Shingle: A composition-type shingle used on a roof and made of a saturated felt paper with ground-up pieces of stone embedded and held in place by asphaltum.

Awl: A tool used to mark wood with a scratch mark; can be used to produce pilot holes.

Awning Window: A window that is hinged at the top and the bottom swings outward.

GLOSSARY *(cont.)*

B

Backfill: Any deleterious material (sand, gravel, and so on) used to fill an excavation.

Backhoe: Self-powered excavation equipment that digs by pulling a boom-mounted bucket toward itself.

Backsplash: The vertical part of a countertop that runs along the wall to prevent splashes from marring the wall.

Balloon Framing: Wall construction extending from the foundation to the roof structure without interruption; used in residential construction only.

Baluster: That part of the staircase that supports the handrail or bannister.

Balustrade: A complete handrail assembly; includes the rails, the balusters, subrails, and fillets.

Bank Plug: Piece of lumber driven into the ground leaving 24" above ground level; surveyors place nails in the bank plugs a given distance above the road surface so a string line can be stretched between the plugs to measure grade.

Bannister: That part of the staircase that fits on top of the balusters.

Baseboard: Molding covering the joint between a finished wall and the floor.

Base Shoe: A molding added at the bottom of a baseboard; used to cover the edge of finish flooring or carpeting.

Batt Insulation: Material formed into sheets or rolls installed between framing members.

Batten: A narrow piece of wood used to cover joints.

Batter Boards: Boards used to frame in the corners of a proposed building while layout/excavation occur.

Beam: A horizontal framing member; may be made of steel or wood; usually refers to a wooden beam at least 5 inches thick and at least 2 inches wider than it is thick.

Bearing Partition: An interior divider or wall that supports the weight of the structure above it.

Bearing Wall: A wall having weight-bearing properties associated with holding up a building's roof or second floor.

Benching: Making steplike cuts into a slope; used for erosion control or to tie a new fill into an existing slope.

Benchmark: Point of known elevation from which the

GLOSSARY *(cont.)*

surveyors can establish all their grades.

Berm: A raised earth embankment; the shoulder of a paved road; the area between the curb and the gutter and a sidewalk.

Bevel: A tool that can be adjusted to any angle; it helps make cuts at the number of degrees that is desired and is a good device for transferring angles from one place to another.

Bibb: A faucet used to connect a hose.

Bi-fold: A double-leaf door used primarily for closet doors.

Bird Mouth: A notch cut into a roof rafter so that it can rest smoothly on the top plate.

Blistering: The condition that paint presents when air or moisture is trapped underneath and makes bubbles that break into flaky particles and ragged edges.

Blocking: A piece of wood fastened between structural members to strengthen them; generally, solid or cross-tie wood or metal cross-tie members.

Board Foot: A unit of lumber measure equaling 144 cubic inches; the base unit (B.F.) is 1 inch thick and 12 inches square or $1 \times 12 \times 12 = 144$ cubic inches.

Bond: In masonry, the interlocking system of brick or block to be installed.

Borrow Site: An area from which earth is taken for hauling to a jobsite that is short of earth needed to build an embankment.

Bottom or Heel Cut: The cutout of the rafter end that rests against the plate; also called the *foot* or *seat cut.*

Bow: A term used to indicate an upward warp along the length of a piece of lumber.

Bow Window: A window unit that projects from an exterior wall.

Brace: An inclined piece of lumber applied to a wall or to roof rafters to add strength.

Bridging: Used to keep joists from twisting or bending.

Builder's Level: A tripod-mounted device that uses optical sighting to make sure that a straight line is sighted and that the reference point is level.

Building Codes: Rules and regulations that are formulated in a code by a local housing authority or governing body.

Building Paper: Also called tar paper, roofing paper, and a number of other terms; paper having a black coating of asphalt for use in weatherproofing.

Building Permits: A series of permits that must be obtained for building; allows for inspections of the work and places house on tax roles.

Butt: To meet edge to edge, such as in a joining of wooden edges.

C

Calcium Chloride: A concrete admixture used for accelerating the cure time.

California Bearing Ratio (CBR): A system used for determining the bearing capacity of a foundation.

Carriage: A notched stair frame.

Casement: A type of window hinged to swing outward.

Casing: The trim that goes on around the edge of a door opening; also the trim around a window.

Catch Basin: A complete drain box made in various depths and sizes; water drains into a pit, then from it through a pipe connected to the box.

Catch Point: Another name for hinge point or top of shoulder.

Caulk: Any type of material used to seal walls, windows, and doors.

Cement: The basis for a concrete mix.

Cement Plaster: A mixture of gypsum, cement, hydrated lime, sand, and water, used primarily for exterior wall finish.

Center Line: The point on stakes or drawings that indicate the halfway point between two sides.

Chair: Small device used to support horizontal rebar prior to the concrete placement.

Chord: Top or bottom member of a truss.

Clear and Grub: To remove all vegetation, trees, concrete, or anything that will interfere with construction.

Cleat: Any strip of material attached to the surface of another material to strengthen, support, or secure a third material.

Collar Tie: Horizontal framing member tying the raftering together above the plate line.

Common Rafter: A structural member that extends without interruption from the ridge to the plate line in a sloped roof structure.

Computer-Aided Drafting (CAD): Computer-aided design and drafting.

Concrete: A mixture of sand, gravel, and cement in water.

Condensation: The process by which moisture in the air becomes water or ice on a surface (such as a window).

Contour Line: Solid or dashed lines showing the elevation of the earth on a project.

Convection: Transfer of heat through the movement of a liquid or gas.

Corner Beads: Metal strips that prevent damage to drywall corners.

Cornice: The area under the roof overhang; usually enclosed or boxed in.

Crawl Space: The area under a floor that is not fully excavated.

Cripple Jack: A jack rafter with a cut that fits in between a hip and a valley rafter.

Cripple Rafter: A cripple rafter is not as long as the regular rafter used to span a given area.

Cripple Stud: A short stud that fills out the position where the stud would have been located if some other opening had not been there.

Cross Brace: Wood or metal diagonal bracing used to aid in structural support between joists and beams.

Crow's Foot: A lath set by the grade setter with markings to indicate the final grade at a certain point.

Cup: To warp across the grain.

Curtain Wall: Exterior walls are called curtain walls if they do not carry loads from roof or floors above them.

Cutting Plane Line: A heavy broken line with arrows, letters, and numbers at each end indicating the section view that is being identified.

D

Dado: A rectangular groove cut into a board across the grain.

Damp-Proofing: Moisture protection; a surfacing used to protect concrete and masonry.

Datum Point: See benchmark; identification of the elevation above mean sea level.

Dead Load: The weight of a structure and all its fixed components.

Deck: The part of a roof that covers the rafters.

Deformed Bar: Steel reinforcement bar with ridges to prevent the bar from

GLOSSARY *(cont.)*

loosening during the concrete curing process.

Diagonal Brace: A wood or metal member placed diagonally over wood or metal framing to add rigidity at corners and at 25'0" feet of unbroken wall space.

Dimension Line: A line on a drawing with a measurement indicating length.

Double Plate: Usually refers to the practice of using two pieces of dimensional lumber for support over the top section or wall section.

Double Trimmer: Double joists used on the sides of openings; double trimmers are placed without regard to regular joist spacings for openings in the floor for stairs or chimneys.

Downspouts: Pipes connected to the gutter to conduct rainwater to ground or sewer.

Drain Tile: Usually made of plastic, generally 4 inches in diameter, with small holes to allow water to drain into it; laid along the foundation footing to drain seepage into a sump or sewer.

Drop Siding: Drop siding has a special groove, or edge cut into it; the edge lets each board fit into the next board and makes the boards fit together to resist moisture.

Ductwork: A system of pipes used to pass heated or cooled air to all parts of a house.

E

Easement: A portion of land on or off a property that is set aside for utility installations.

Eave: The lowest edge on a gable roof.

Eaves: The overhang of a roof projecting over the walls.

Eaves Trough: A gutter.

Elevation: An exterior or interior orthographic view of a structure identifying the design and the materials.

Elevation Numbers: The vertical distance above or below sea level.

Embankment: Area being filled with earth.

F

Face: The exposed side of a framing or masonry unit; a type of brick; also called *common.*

Fascia: A flat board covering the ends of rafters on the cornice or eaves where gutters are mounted.

Feathering: Raking new asphalt to join smoothly with the existing asphalt.

GLOSSARY *(cont.)*

Finish: Any material used to complete an installation that provides an aesthetic or finished appearance.

Firebrick: A special type of brick that is not damaged by fire; used to line the firebox in a fireplace.

Fire Stop/Draft Stop/Fire Blocking: A framing member used to reduce the ability of a fire's spread.

Firewall/Fire Separation Wall/ Fire Division Wall: Any wall that is installed for the purpose of preventing the spread of fire.

Flashing: Metal or plastic strips or sheets used for moisture protection in conjunction with other construction materials.

Flat: In roofing, any roof structure up to a 3:12 slope.

Flue: The passage through a chimney.

Flush: To be even with.

Fly Ash: Fine, powdery coal residue used with a hydraulic (water-resistant) concrete mix.

Footing: The bottom-most member of a foundation; supports the full load of the structure above.

Form: A temporary construction member used to hold permanent materials in place.

Foundation: The base on which a house or building rests.

Frostline: The depth to which ground freezes in the winter.

Furring Strips: Strips of wood attached to concrete or stone that form a nail base for wood or paneling.

G

Gable: The simplest kind of roof; two large surfaces come together at a common edge forming an inverted V.

Gambrel Roof: A barn-shaped roof.

Gauge: The thickness of metal or glass sheet material.

Girder: A support for joists at one end; usually placed halfway between the outside walls and runs the length of the building.

Glaze: To install glass.

Glu-Lam (GLB): Glue-laminated beam made from milled 2x lumber bonded together to form a beam.

Grade: An existing or finished elevation in earthwork; a sloped portion of a roadway; sizing of gravel and sand; the structural classification of lumber.

Grade Break: A change in slope from one incline ratio to another.

GLOSSARY *(cont.)*

Grade Lath: A piece of lath that the surveyor or grade setter has marked to indicate the correct grade to the operators.

Grade Pins: steel rods driven into the ground at each surveyor's hub; a string is stretched between them at the grade indicated on the survey stakes, or a constant distance above the grade.

Grader: A power excavating machine with a central blade that can be angled to cast soil on either side.

Gravel Stop: The edge metal used at the eaves of a built-up roof to hold the gravel.

Green: Uncured or set concrete or masonry; freshly cut lumber.

Grid System: A system of metal strips that support a drop ceiling.

Ground-Fault-Circuit-Interrupter (GFCI) or (GFI): An electrical receptacle installed for personal safety at outdoor locations and near water.

Guinea: A survey marker driven to grade; it may be colored with paint or crayon; used for finishing and fine trimming; also called a *hub.*

Gusset: A triangular or rectangular piece of wood or metal that is usually fastened to the joint of a truss to strengthen it; used primarily in making roof trusses.

Gutter: A metal or wooden trough set below the eaves to catch and conduct water to a downspout.

Gypsum: A chalk used to make wallboard; produces a plastered wall with certain fire-resisting characteristics.

H

Habitable Space: In residential construction, the interior areas of a residence used for eating, sleeping, living, and cooking; excludes bathrooms, storage utility rooms, etc.

Hanger: Metal fabrication made for the purpose of placing and supporting joists and rafters.

Hardware: Any component used to hang, support, or position another component; e.g., doors and windows.

Hardwood: Wood that comes from trees that shed leaves.

Head Joint: The end face of a brick or block to which mortar is applied.

Header: A framing member used to hide the ends of joists along the perimeter of a structure; also known as a *rim joist*; the horizontal structural framing member installed over wall openings to aid in the support of the structure above; also referred to as a *lintel.*

Header Course: In masonry, a horizontal row of brick laid perpendicular to the wall face; used to tie a double wythe brick wall together.

Hidden Line: A dashed line identifying portions of construction that are a part of the drawing but cannot be seen; e.g., footings on foundation plans, etc.

Hip Rafters: Extend diagonally from the plate to the ridge.

Hip Roof: A structural sloped roof design with sloped perimeter from ridge to plate line.

Hose Bibb: A faucet used to connect a hose.

HVAC: Heating, Ventilating and Air Conditioning. The mechanical portion of the CSI format, division 15.

Hydraulic Cement: A cement capable of curing under water.

I

Insulation: Any material capable of resisting thermal, sound, or electrical transmission.

Insulation Resistance: The R factor in insulation calculations.

J

Jack Rafter: A part of the roof structure raftering that does not extend the full length from the ridge beam to the top plate.

Jamb: The part that surrounds a door or window frame; usually made of two vertical pieces and a horizontal piece over the top.

Joint Compound: Material used with a paper of fiber tape for sealing indentations and breaks in drywall construction.

Joist: A structural horizontal framing member used for floor and ceiling support.

Joist Hangers: Metal brackets that hold up the joist; they are nailed to the girder.

K

Key: A depression made in a footing so that the foundation or wall can be poured into the footing, preventing the wall or foundation from moving during changes in temperature or settling of the building.

Kicker Blocks: Cement poured behind each bend or angle of water pipe for support; also called *thrust blocks.*

Kiln-Dried Lumber: Lumber that is seasoned under controlled conditions, removing from 6% to 12% of the moisture.

King Stud: A full-length stud from bottom plate to the top plate supporting both sides of a wall opening.

Knee Wall: Vertical framing members supporting and shortening the span of the roof rafters.

L

Lateral: Underground electrical service.

Lath: Backup support for plaster; may be of wood, metal, or gypsum board.

Lavatory: Bathroom; basin.

Lay-in Ceiling: A suspended ceiling system.

Leach Line: A perforated pipe used as a part of a septic system to allow liquid overflow to dissipate into the soil.

Ledger: Structural framing member used to support ceiling and roof joists at the perimeter walls.

Level-Transit: An optical device that is a combination of a level and a means for checking vertical and horizontal angles.

Lift: Any layer of material or soil placed upon another.

Live Load: Any movable equipment or personal weight to which a structure is subjected.

Load: The weight of a building.

Load Conditions: The conditions under which a roof must perform.

Lockset: The doorknob and associated locking parts inserted in a door.

M

Masonry: Manufactured materials of clay (brick), concrete (CMU), and stone.

Mastic: An adhesive used to hold tiles in place; also refers to adhesives used in construction.

Mat: Asphalt as it comes out of a spreader box or paving machine in a smooth, flat form.

Maximum Density and Optimum Moisture: The highest point on the moisture density curve; considered the best compaction of the soil.

Membrane Roofing: Built-up roofing.

Mesh: Common term for welded wire fabric, plaster lath.

Mil: 0.001".

Minute: 1/60th of a degree.

Moisture Barrier: A material used for the purpose of resisting exterior moisture penetration.

GLOSSARY *(cont.)*

Moisture Density Curve: A graph plotted from tests to determine at what point of added moisture the maximum density will occur.

Moldings: Trim mounted around windows, floors, and doors as well as closets.

Monolithic Concrete: Concrete placed as a single unit including turndown footings.

Mortar: A concrete mix especially used for bonding masonry units.

N

Natural Grade: Existing or original grade elevation.

Natural Ground: The original ground elevation.

Nominal Size: Original cut size of a piece of lumber prior to milling and drying; size of masonry unit, including mortar bed and head joint.

Nonbearing: Not supporting any structural load.

Nuclear Test: A test to determine soil compaction by sending nuclear impulses into the compacted soil and measuring the returned impulses.

O

On Center (O/C): The distance between the centers of two adjacent components.

Open Web Joist: Roof joist made of wood or steel construction with a top chord and bottom chord connected by diagonal braces bolted or welded together.

P

Pad: In earthwork or concrete foundation work, the base materials used upon which to place the concrete footing and/or slab.

Parapet: An extension of an exterior wall above the line of the roof.

Parging: A thin moisture-protection coating of plaster or mortar over a masonry wall.

Partition: An interior wall separating two rooms or areas of building; usually nonbearing.

Penny (d): The unit of measure of the nails used by carpenters.

Perimeter: The outside edges of a plot of land or building; it represents the sum of all the individual sides.

Perimeter Insulation: Insulation placed around the outside edges of a slab.

Pile: A steel or wooden pole driven into the ground sufficiently to support the weight of a wall and building.

GLOSSARY *(cont.)*

Pillar: A pole or reinforced wall section used to support the floor and consequently the building.

Pitch: The slant or slope from the ridge to the plate.

Plan View: A bird's-eye view of a construction layout cut at 5'0" above finish floor level.

Plate: A roof member that has the rafters fastened to it at their lower ends.

Platform Framing: Also known as *western framing*; structural construction in which all studs are only one-story high with joists over.

Point of Beginning (POB): The point on a property from which all measurements and azimuths are established.

Polyvinyl Chloride (PVC): Plastic pipe commonly used for plumbing.

Portland Cement: One variety of cement produced from burning various materials such as clay, shale, and limestone, producing a fine gray powder; the basis of concrete and mortar.

Post-and-Beam Construction: A type of wood frame construction using timber for the structural support.

Post-Tensioning: The application of stretching steel cables embedded in a concrete slab to aid in strengthening the concrete.

Prehung: Refers to doors or windows that are already mounted in a frame and are ready for installation.

Pressure Treatment: Impregnating lumber with a chemical under pressure.

Primer: The first coat of paint or glue when more than one coat will be applied.

Purlin: A horizontal framing member spanning between rafters.

Q

Quarry Tile: An unglazed clay or shale flooring material.

Quick Set: A fast-curing cement plaster.

R

R Factor: The numerical rating given any material that is able to resist heat transfer for a specific period of time.

R Values: The unit that measures the effectiveness of insulation; the higher the number, the better the insulation qualities.

Rabbet: A groove cut in or near the edge of a piece of lumber to fit the edge of another piece.

GLOSSARY *(cont.)*

Raceway: Any partially or totally enclosed container for placing electrical wires.

Rafter: The framing member extending from the ridge or hip to the top plate.

Rebar: A reinforcement steel rod in a concrete footing.

Resilient Flooring: Flooring made of plastics rather than wood products.

Ridge: The highest point on a sloped roof.

Ridge Board: A horizontal member that connects the upper ends of the rafters on one side to the rafters on the opposite side.

Right-of-Way Line: A line on the side of a road marking the limit of the construction area and usually, the beginning of private property.

Rise: In roofing, rise is the vertical distance between the top of the double plate and the center of the ridge board; in stairs, it is the vertical distance from the top of a stair tread to the top of the next tread.

Riser: The vertical part at the edge of a stair.

Roll Roofing: A type of built-up roofing material made of a mixture of rag, paper, and asphalt.

Roof Pitch: The ratio of total span to total rise expressed as a fraction.

Rough Opening: A large opening made in a wall frame or roof frame to allow the insertion of a door or window.

RS: Reference stake, from which measurements and grades are established.

Run: The run of a roof is the shortest horizontal distance measured from a plumb line through the center of the ridge to the outer edge of the plate.

S

Sand Cone Test: A test for determining the compaction level of soil.

Scabs: Boards used to join the ends of a girder.

Schematic: A one-line drawing for electrical circuitry or isometric plumbing diagrams.

Scissors Truss: A truss constructed to the roof slope at the top chord with the bottom chord designed with a lower slope for interior vaulted or cathedral ceilings.

Scraper: A digging, hauling, and grading machine.

Scratch Coat: First coat of plaster placed over lath in a three-coat plaster system.

GLOSSARY *(cont.)*

Scupper: An opening in a parapet wall attached to a downspout for water drainage.

Scuttle: Attic or roof access with cover or door.

Sealant: A material used to seal off openings against moisture and air penetration.

Section: A vertical drawing showing architectural or structural interior design developed at the point of a cutting-plane line on a plan view; the section may be transverse—the gable end—or longitudinal—parallel to the ridge.

Seismic Design: Construction to withstand earthquakes.

Septic System: A waste system that includes a line from the structure to a tank and a leach field.

Shakes: Shingles made of handsplit wood; in most cases western cedar.

Shear Wall: A wall construction designed to withstand shear pressure caused by wind, etc.

Sheathing: The outside layer of wood applied to studs to close up a house or wall; also used to cover the rafters and make a base for the roofing.

Sheepsfoot Roller: A soil compacting roller with feet expanded at their outer tips.

Shoring: Temporary support made of metal or wood.

Sill: A piece of wood that is anchored to the foundation.

Sinker Nail: A nail for laying subflooring; the head is sloped toward the shank but is flat on top.

Size: A special coating used for walls before wallpaper is applied; it seals the walls and allows the wallpaper paste to attach itself to the wall and paper.

Slab-on-Grade: A foundation for a structure with no crawl space or basement.

Slump: The consistency of concrete when poured.

Soffit: A covering for the underside of the overhang of a roof.

Soleplate: A 2 × 4 or 2 × 6 used to support studs in a horizontal position; it is placed against the flooring and nailed into position onto the subflooring.

Span: The horizontal distance between exterior bearing walls in a transverse section.

Spoil Site: Area used to dispose of unsuitable or excess excavation material.

Spreader: Braces used across the top of concrete forms.

GLOSSARY *(cont.)*

Square: Refers to a roof-covering area; a square consists of 100 square feet of area.

Stain: A paint-like material that imparts a color to wood.

Stepped Footing: A footing that may be located on a number of levels.

Stool: The flat shelf that rims the bottom of a window frame on the inside of a wall.

Stress Skin Panels: Large prebuilt panels used as walls, floors, and roof decks.

String Line: A nylon line usually strung tightly between supports to indicate both direction and elevation; used in checking grades or deviations in slopes or rises.

Strip Flooring: Wooden strips that are applied perpendicular to the joists.

Strongbacks: Braces used across ceiling joints that help align, space, and strengthen joists for drywall installation.

Structural Steel: Heavy steel members larger than 12 gauge.

Stucco: A type of finish used on the outside of a building; a masonry finish that can be put on over any type of wall.

Stud: The vertical boards (usually 2 × 4 or 2 × 6) that make up the walls of a building.

Subfloor: A platform that supports the rest of the structure; also referred to as underlayment.

Subgrade: The uppermost level of material placed in embankment or left at cuts in the grading of a road bed.

Summit: The highest point of any area or grade.

Super: A continuous slope in one direction on a road.

Swale: A shallow dip made to allow the passage of water.

Sway Brace: A piece of 2 × 4 or similar material used to temporarily brace a wall from wind until it is secured.

Swedes: A method of setting grades at a center point by sighting across the tops of three lath; two lath are placed at a known correct elevation and the third is adjusted until it is at the correct elevation.

Symbol: A pictorial representation of a material or component on a plan.

T

Tail/Rafter Tail: That portion of a roof rafter extending beyond the plate line.

Tail Joist: A short beam or joist supported in a wall on one end and by a header on the other.

GLOSSARY *(cont.)*

Tamp: To pack tightly; usually refers to making sand tightly packed or making concrete mixed properly in a form to get rid of air pockets.

Tangent: A straight line from one point to another, which passes over the edge of a curve.

Taping and Bedding: Refers to drywall finishing; taping is the application of a strip of specially prepared tape to drywall joints; bedding means embedding the tape in the joint to increase its structural strength.

Tensile Strength: The maximum stretching of a piece of metal before breaking.

Tensioning: Pulling or stretching of steel tendons to reinforce concrete.

Terrazzo: A mixture of concrete, crushed stone, calcium shells, and/or glass, polished to form a tile-like finish.

Texture Paint: A very thick paint that will leave a texture or pattern.

Thermal Ceilings: Ceilings that are insulated with batts of insulation.

Tie: A soft metal wire that is twisted around a rebar or reinforcement rod and chair to hold the rod in place until concrete is poured.

Tied Out: The process of determining the fixed location of existing objects (manholes, meter boxes, etc.) in a street so that they may be uncovered and raised after paving.

Toe of Slope: The bottom of an incline.

Top Chord: The topmost member of a truss.

Top Plate: The horizontal framing member fastened to the top of the wall studs; usually doubled.

Track Loader: Used for filling, and loading materials.

Tread: The part of a stair on which people step.

Trimmer: A piece of lumber, usually a 2 × 4, that is shorter than the stud or rafter but is used to fill in where the longer piece would have been normally spaced except for some opening in the roof or floor or wall.

Truss: A prefabricated sloped roof system incorporating a top chord, bottom chord, and bracing.

U

Underlayment: Also referred to as the subfloor; used to

support the rest of the building; may also refer to the sheathing used to cover rafters and serve as a base for roofing.

Unfaced Insulation: Insulation that does not have a facing or plastic membrane over one side of it; it has to be placed on top of existing insulation; if used in a wall, it has to be covered by a plastic film to ensure a vapor barrier.

V

Valley: The area of a roof where two sections come together and form a depression.

Valley Rafters: A rafter that extends diagonally from the plate to the ridge at the line of intersection of two roof surfaces.

Vapor Barrier: The same as a moisture barrier.

Veneer: A thin layer or sheet of wood.

Vent: Usually a hole in the eaves or soffit to allow the circulation of air over an insulated ceiling; usually covered with a piece of metal or screen.

Vent Stack: A system of pipes used for air circulation and to prevent water from being suctioned from the traps in the waste disposal system.

Ventilation: The exchange of air, or the movement of air through a building; may be done naturally through doors and windows or mechanically by motor-driven fans.

Vibratory Roller: A self-powered or towed compacting device.

W

Waler: A 2x piece of lumber installed horizontally to formwork to give added stability and strength to the forms.

Water-Cement Ratio: The ratio between the weight of water to cement.

Waterproofing: Preferably called moisture protection; materials used to protect below- and on-grade construction from moisture penetration.

Water Table: The amount of water that is present in any area.

Welded Wire Fabric (WWF): A reinforcement used for horizontal concrete strengthening.

Wind Lift (Wind Load): The force exerted by the wind against a structure; caused by the movement of the air.

Winder: Fan-shaped steps that allow the stairway to

change direction without a landing.

Window Apron: The flat part of the interior trim of a window; located next to the wall and directly beneath the window stool.

Window Stool: The flat narrow shelf that forms the top member of the interior trim at the bottom of a window.

Windrow: The spill-off from the ends of a dozer or grader blade that forms a ridge of loose material; a windrow may be deliberately placed for spreading by another machine.

Wythe: A continuous masonry wall width.

XYZ

X Brace: Cross brace for joist construction.

CHAPTER 13
Abbreviations and Symbols

ABBREVIATIONS

A	area
AB	anchor bolt
AC	alternate current
A/C	air conditioning
ACT	acoustical ceiling tile
A.F.F.	above finish floor
AGGR	aggregate
AIA	American Institute of Architects
AL, ALUM	aluminum
AMP	ampere
APPROX	approximate
ASPH	asphalt
ASTM	American Society for Testing Materials
AWG	American Wire Gauge
BD	board
BD FT (BF)	board foot (feet)
BLDG	building
BLK	black, block
BLKG	blocking
BM	board measure
CC	center to center, cubic centimeter
CEM	cement
CER	ceramic
CFM	cubic feet per minute
CIP	cast-in-place, concrete-in-place
CJ	ceiling joist, control joint
CKT	circuit (electrical)
CLG	ceiling
CMU	concrete masonry unit
CO	cleanout (plumbing)
COL	column
CONC	concrete
CONST	construction
CONTR	contractor
CU FT (ft^3)	cubic foot (feet)
CU IN (in^3)	cubic inch(es)
CU YD (yd^3)	cubic yard(s)
d	pennyweight (nail)
DC	direct current (elec.)
DET	detail
DIA	diameter
DIAG	diagonal
DIM	dimension
DN	down
DO	ditto (same as)
DS	downspout
DWG	drawing
E	east
EA	each
ELEC	electric, electrical

ABBREVIATIONS *(cont.)*

ELEV	elevation, elevator
ENCL	enclosure
EXCAV	excavate, excavation
EXT	exterior
FDN	foundation
FIN	finish
FIN FLR	finish floor
FIN GRD	finish grade
FL, FLR	floor
FLG	flooring
FLUOR	fluorescent
FOB	free-on-board, factory-on-board
FOM	face of masonry
FOS	face of stud, flush on slab
FT	foot, feet
FTG	footing
FURN	furnishing, furnace
FX GL (FX)	fixed glass
GA	gauge
GAL	gallon
GALV	galvanize(d)
GD	ground (earth/electric)
GI	galvanized iron
GL	glass
GL BLK	glass block
GLB, GLU-LAM	glue-laminated beam
GRD	grade, ground
GWB	gypsum wall board
GYP	gypsum
HB	hose bibb
HDR	header
HDW	hardware
HGT/HT	height
HM	hollow metal
HORIZ	horizontal
HP	horsepower
HWH	hot water heater
ID	inside diameter
IN	inch(es)
INSUL	insulation
INT	interior
J, JST	joist
JT	joint
KG	kilogram
KL	kiloliter
KM	kilometer
KWH	kilowatt-hour
L	left, line
LAU	laundry
LAV	lavatory
LBR	labor
LDG	landing, leading
LDR	leader
LEV/LVL	level
LIN FT (LF)	lineal foot (feet)
LGTH	length
LH	left hand
LITE/LT	light (window pane)
MAT'L	material
MAX	maximum

ABBREVIATIONS *(cont.)*

MBF/MBM	thousand board feet, thousand board measure
MECH	mechanical
MISC	miscellaneous
MK	mark (identifier)
MO	momentary (electrical contact), masonry opening
N	north
NEC	National Electric Code
NIC	not in contract
NOM	nominal
O/A	overall (measure)
O.C. (O/C)	on center
OD	outside diameter
OH	overhead
O/H	overhang (eave line)
OPG	opening
OPP	opposite
PC	piece
PLAS	plastic
PLAST	plaster
PLT	plate (framing)
PR	pair
PREFAB	prefabricate(d)(tion)
PTN	partition
PVC	polyvinylchloride pipe
QT	quart
QTY	quantity
R	right
RD	road, round, roof drain
REBAR	reinforced steel bar
RECEPT	receptacle
REINF	reinforce(ment)
REQ'D	required
RET	retain(ing), return
RF	roof
RFG	roofing (materials)
RH	right hand
S	south
SCH/SCHED	schedule
SECT	section
SERV	service (utility)
SEW	sewer
SHTHG	sheathing
SIM	similar
SP	soil pipe (plumbing)
SPEC	specification
SQ FT (ft^2)	square foot (feet)
SQ IN (in^2)	inch(es)
SQ YD (yd^2)	square yard(s)
STA	station
STD	standard
STIR	stirrup (rebar)
STL	steel
STR/ST	street
STRUCT	structural
SUSP CLG	suspended ceiling
SYM	symbol, symmetric
SYS	system

ABBREVIATIONS *(cont.)*

T&G	tongue and groove	**W**	west
THK	thick	**w/**	with
TOB	top of beam	**w/o**	without
TOC	top of curb	**WC**	water closet (toilet)
TOF	top of footing	**WDW**	window
TOL	top of ledger	**WI**	wrought iron
TOP	top of parapet	**WP**	waterproof, weatherproof
TOS	top of steel		
TR	tread, transition	**WT/WGT**	weight
TRK	track, truck		
TYP	typical	**YD**	yard
UF	underground feeder (electrical)	**Z**	zinc
USE	underground service entrance cable (electrical)		

SYMBOLS

&	and	"	ditto, inch, -es
∠	angle	'	foot, feet
@	at	%	percent
#	number, pound	Ø	diameter

HOUSEHOLD FIXTURES

Elevation Views

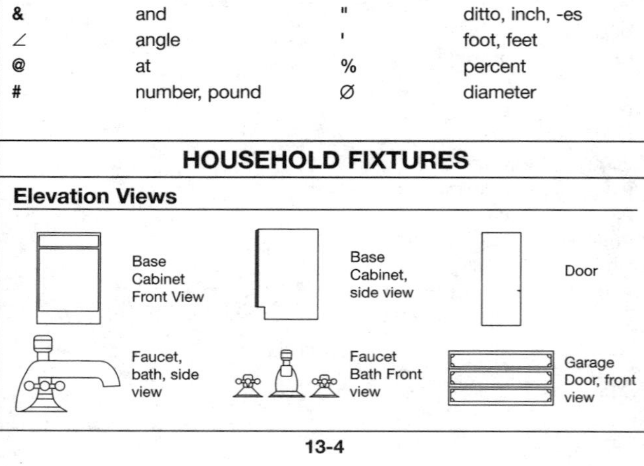